THE CRAFT MALTSTERS' HANDBOOK

BY DAVE THOMAS

WHITE MULE PRESS
NICHE BOOKS FOR LOVERS OF SPIRITS

© 2014 by White Mule Press a division of the
American Distilling Institute™. All rights reserved.

Printed in the United States of America.

ISBN 978-0-9910436-2-0

Cover illustration courtesy of Special Collections,
University of Amsterdam, inv. 076.375.

TABLE OF CONTENTS

INTRODUCTION	1		
A	5	L	104
B	12	M	105
C	32	N	131
D	52	O	133
E	61	P	134
F	65	Q, R	143
G	77	S	147
H	95	T	171
I	98	U, V	174
K	99	W	175

REFERENCES 182

APPENDIX A 188
The malting process flow of heat, air, grain and water

APPENDIX B 190
Recommended Malt Characteristics
American Malting Barley Association

APPENDIX C 192
North American Craft Malting Companies

APPENDIX D 196
Barley to Beverage in a Ballad

FOREWORD

There is an old saying that if the student does not do better than the teacher then it is the teacher who has failed. When I was asked to write the foreword for this wonderful handbook on malting, it was comforting to know that another one of my students had done better than me.

Dave Thomas arrived at the Heriot-Watt Brewing School, Edinburgh, many years ago, from Coors, to study successfully for his MSc under my supervision. Dave's handbook contains up-to-date references of the many different types of malts used in brewing today. However, he also quotes Burns on barley (1782) and Stopes on steeping (1885) in this excellent book on the technical and scientific meanings of words and phrases used in malting. In passing, Burns not only wrote one of the most meaningful and popular songs in the world, *Auld Lang Syne*, which is sung to herald in the New Year, he also grew malting barley for a living, which was used to make Scottish Ale and Scotch Whisky.

In this handbook, words and their meanings are laid out alphabetically. For example, words in the A section contains: Abrasion, Acrospire, Aeration, Aleurone layer, Amber malt, American Malting Barley Association, American Society of Brewing Chemists, Amylolytic (starch-degrading) enzymes, Apparent Extract, and Assortment (grain sizing or percent plumpness).

Thumbing through the D and the M sections one finds: Diastase, Diastatic power, DON, Drum roaster, Drying barley, Dust explosion: Malt damage, Malt extract, Malt evaluation, Malting, Malt analysis, Maris Otter, Mash conversion iodine test, Master Brewers' Association of the Americas, Middle European Brewing Technology Analysis Commission, Melanoidins, Micronized grains, Microscopic Evaluation of malts, Millet, and Milling.

A glance at the S section reveals: Slack malt, Smut, Sorghum, Speciality Malts, Spelt, Spent grains, Spirit yield, Spring barley, Starch, and Steeping. Words reflecting a wide range of topics such as Storage, Microbes, Germination,

Kilning, Malt storage, Mashing, Brewing, and Distilling are also covered.

The mention of Maris Otter in the book reminded me that one of Maris Otter's parents was Proctor, the super malting barley, whose endosperm cell walls were less rigid, released less viscous beta-glucans, and were more susceptible to enzymic and structural breakdown than the corresponding cell walls of the poor quality barley Julia. Proctor and Maris Otter were selected by Dr. Bell of the Plant Breeding Institute at Cambridge, UK. When I asked him, about 40 years ago, how he selected these super malting barley varieties, he simply replied: "By eye."

This is a handbook produced with much love and years of experience that will be of great value to a wide variety of people who work in or who have an interest in this exciting industry. The book describes the importance of malt in many ways, but to assess the importance of malt in our history, one only has to look at Article 13 of the political Union of 1707 that made Scotland a part of a Great Britain "that ruled the waves." Article 13 was the last of the 25 Articles to be signed. When it was eventually signed in 1725 there were fatalities in the street riots that followed, in Glasgow. What was Article 13 about? It was about the new tax on malt, which was judged to be unfair.

Professor Sir Geoffrey Palmer
Penicuik, Scotland
April, 2014.

ACKNOWLEDGEMENTS

Heartfelt thanks goes out to some of the very people who, today, are rapidly creating, learning and teaching each other the revitalized art of craft malting including Brandon Ade, Wendell Banks, the Cody family (Jason, Josh, Tim, and Wayne), Jim Eckert, Twila Henley, Chris Schooley, Andrea and Christian Stanley, and Bruno Vachon; my co-workers and friends in the malting, brewing, and distilling industries, Mike Davis, Kristen French, Stephen Gould, Scott Heisel, Joe Hertrich, Ian Hornsey, Mike Joyce, Rob McCaig, Bill Owens, Brad Plummer, Gail Sands, Roger Putman, Tom Shellhammer, Paul Schwarz, and Mont Stuart; and to my malting and brewing professor, mentor, editor and friend, Professor Sir Geoffrey Palmer.

Without my wife Amy's infinite patience and grandson Jameson's permission to "use my bedroom to do your homework," I never would have finished!

INTRODUCTION

Malt has often been called the backbone, workhorse, magic, soul, and romance of beer, distilled spirits, and foods. After decades of "hop happiness" in the U.S. craft beer industry, craft malt is fast becoming the essential, enigmatic ingredient that fills craft beer, craft foods, and craft spirits with color, flavor, strength, personality, and "terroir" (a sense of place).

Attempting to join with the 20% increase in craft beer volume in 2013 (2,822 breweries adding $14.3 billion to the U.S. economy) and the 25% increase in craft distilling (nearing 700 U.S. craft distilleries), the interest in craft malting has recently taken off in the U.S., growing from fewer than five craft malting companies in 2010, to more than 40 in 2014.

The art and science of malting is many hundreds of years old, but still being examined closely today. Woe to the maltster, brewer, or distiller who simply treats malt as a commodity ingredient. It is wise to use all six senses—sight, smell, hearing, feeling, taste, and common sense—in making and using malt.

The industrial process of germination—which converts hard, insoluble cereals into friable, extractable grains for subsequent use as a food source for humans or yeast—is called malting. Cereal grains are members of the grass families *Gramineae* and *Poaceae*, which are of commercial interest as food, being a rich source of vitamins, minerals, fats, oils, proteins, and carbohydrates.

The primary reason for malting cereals is to enzymically modify a grain's endosperm food reserves so the expected brewer's extract or distiller's spirit yield can be achieved. Modified reserves such as starch are then converted into a sugary extract during mashing. These sugars are fermented into the ethanol in beers and whiskies.

The cereal grain used most often in malting is barley (*Hordeum vulgare*), but wheat, sorghum, rye, oats, amaranth, buckwheat, triticale, rice, emmer, einkorn, millet, quinoa, teff, spelt, and maize can also be malted.

Malting of the "wild" barley cultivar *Hordeum spontaneum* as a food source started in Neolithic times. Accidental microbial infection of germinated cereal gruel most likely produced the first fermented beverage. This fortuitous production of fermented beverages ultimately evolved into the art, science, technological, and business disciplines of malting, brewing, and distilling as they are today.

Most malthouses in the 19th century began as attachments to breweries and distilleries to ensure their own malt supply. In 1833, there were 13,242 official malting licenses taken out in the UK for this purpose (Brown, 1983). Since then, worldwide consolidation has occurred in the malting business, as all others, so there are far fewer brewers and distillers making their own malt, being replaced by a few dozen multinational sales malting companies. Today, the top 20 worldwide malting companies are all sales maltsters, producing an average of five million pounds of malt daily.

Brewery and distillery malthouses have the advantage of a committed demand for most of their product and many of them also sell surplus malt to others. Farm malthouses can provide all or a large part of their raw grain resources by planting it themselves. Sales maltsters usually reserve their grain supplies and predict malt customer volumes through proactive contracts at both ends of the supply chain. Commission maltsters are sales maltsters that malt a brewery or distillery's working capital of grain, according to their specifications.

The role of maltsters, brewers, and distillers was described by H. E. Armstrong (1934) as one of "…fashioning into a scientifically controlled industry, the noble art of converting barley-malt into a wort upon which yeast may feast and ultimately fatten into a beverage with character."

The measure of success of producing a "beverage with character" is, ultimately, known only to the consumer, but the measures of success in converting cereal grains into malt are better understood and are collectively known as malt modification.

Malt modification is the direct result of the biochemical

relationships between the enzymes and food reserves in the steeped (hydrated) endosperm of cereal grains. The same relationships in the agriculture of cereal are measured by the plant's botanical efficiency to survive, grow and propagate within one growing season. Neither malting nor agriculture would be possible within the time limits involved, without proper hydration, specialized structural and chemical characteristics of the grain cell walls and food reserves, and the aid of catalysts such as enzymes that greatly enhance the rate of chemical reactions.

After proper hydration, and (assuming 100% biological viability of grains and oxygen availability) the success of malting or plant growth is dependent on the availability, activity, and distribution of enzymes within the cereal grain, coupled with the susceptibility of the endosperm cell walls and proteins to enzymatic break down or hydrolysis.

The present knowledge of the biochemistry of malting is a cumulative result of work by generations of research scientists and practical maltsters and has given rise to a significant increase in the ability of maltsters to manipulate the malting process to their technological and economic advantage. As an example, a little over 100 years ago, barley malting required from 15 to 31 days, with malting losses of 19 to 24% (Stopes, 1885). Modern day maltsters, however, are able to produce better quality malt in 5 to 7 days and have reduced malting losses to 4 to 10% dry basis, using new malting technologies and improved varieties of barley.

Research into malting biochemistry has often progressed along specific, narrow lines within seemingly "pure research" or academic interest. Examples of these may be separation and characterization of highly specific enzymes and substrates, in vitro microscopic examination of cereal grain tissues, or genetic manipulation of agronomic and malting characteristics. However, all of this information can be used by the practical maltster, brewer, and distiller who, equipped with an understanding of the relationships between hydration, food reserves, and enzymes within the cereal grain, can then manipulate cereal grains and their products into efficient use in the brewhouse, distillery, or food products company.

The present volume endeavors to catalogue many of these historic, scientific, and contemporary aspects of cereal grain malting in a practical handbook format with alphabetical listings for ease of searching and utility by the practicing or aspiring maltster. When a term is cross-referenced in another definition title it is shown in **bold type**.

Appendix A shows a process flow diagram for traditional malting. Appendix B is a table reflecting the recommendations of malt characteristics from the American Malting Barley Association. Appendix C contains a dynamic list of craft malting companies in North America. This list is growing so rapidly that it was updated with new craft malthouses many times during the last few weeks of writing this handbook. There is so much recent interest in this revived hand-crafted industry that maintaining an up-to-date listing is nearly impossible.

Consider the present volume a snapshot in time displaying some of the historic, current, and future prospects for the art, science, and magic of malting on a small scale!

> "Therefore if you have Wisdom and understanding of Nature, remember, that the nearer you come to Nature, and the more you imitate her, the nearer you are to the Truth."   *Tryon (1691).*

A

ABRASION Abrasion is the mechanical abrasion of barley husk that facilitates rapid uptake of steep water, oxygen, and exogenous gibberellic acid, reducing grain moisture and shortening steeping and germination times. The Simon impact abrader consisted of rotating discs, steel pins and internal abrasive surface against which the grains impacted and could process 10 tonnes of barley per hour with 0.5 to 1.0% grain loss. The Sizer abrader used paddles and tapered cylinders and reported 0.3–0.5% grain loss at similar rates as the Simon system.

ABSCISIC ACID A plant hormone involved in stress tolerance, germination, and dormancy. The hormonal balance between abscisic and gibberellic acids determines, *inter alia*, malting grain's germinative energy, capacity, and water sensitivity.

ACID, OR ACIDULATED MALTS Malts that contain lactic acid arising from the growth of lactic acid-producing bacteria or by adding lactic acid directly to steep water.

Kiln-dried malts, when steeped in 113–122°F (45–50°C) water overnight can acidify due to the lactic bacteria growing on the surface of the malt. The acid malt contains as much as 1% lactic acid at this point and is gently dried at 122–140°F (50–60°C) for final malt colors of 2–3 SRM (4–6 EBC).

Acid malts comply with **Rheinheitsgebot** rules, and are added at 3–10% of the brewing grist to reduce wort pH (optimum 5.2 to 5.6) by about 0.1 for every 1% of acid malt added.

Inoculating grain during germination with lactic acid-producing bacteria, like *Lactobacillus amylovorus*, or adding them to the mash, can amplify the acidification process. Diluted sulphuric acid steeps have also been used to inhibit rootlet growth, thereby reducing malting loss.

ACRYLAMIDE NITRILE A carcinogen important to the distilling and food industries. Formed from a reaction between asparagine and reducing sugars in kilned malt.

ACROSPIRE A part of the barley **embryo**, the acrospire, also called the shoot or barley coleoptile, grows under the husk during germination. With varietal differences and practice by the maltster, the length of acrospire growth can be used to estimate degree of internal malt **modification**.

It is desirable for maltsters to control acrospire growth so that it does not extend past the distal end of the kernel, in many cereals. Green malt with overgrown acrospires, also called "overs" or "hussars" can cause higher malt loss when the malt is cleaned.

For assessing barley growth and modification during malting, it is desirable for the acrospire to reach three-quarters to a full length of the barley kernel, with less than 15% overgrowth. Mean acrospire growth relative to the length of the kernel, can easily be counted on 100 kernels and recorded during the final 24 hours of germination.

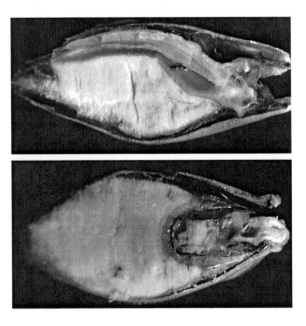

Fig. 1 (pg 6, above): Germinated barley dissected longitudinally showing acrospire growth to three-quarters kernel length, white "mealy" modified endosperm, and dark "stealy" unmodified hard distal end. Fig. 2 (pg 6, below): Transverse dissection showing cut acrospire, embryo, and endosperm.

AERATION To dissolve air in a liquid. Immersion steep waters contain a small percentage of oxygen that is depleted rapidly, in as little as one hour, by barley during steep soak cycles. Continuously bubbling air through steep water (rousing), while not replenishing all of the oxygen in water, can help cool and lift grains and particulates, and promote even germination.

Fig. 3: Vigorous compressed air rousing during steep soak (immersion).

ALEURONE LAYER, ALSO CALLED THE SHOOT The outer layer of living tissue of the barley grain constituting 12% of kernel weight composed of **pentosans, protein, lipid**, and phytic acid. The aleurone layer is the main tissue for production and secretion of enzymes into the endosperm and is the main constituent of the bran fraction of milled cereals. These include the primary starch (amylolytic) enzyme **α-amylase** as well as **ß-glucanase** and **proteases**. **Gibberellic acid** and other plant hormones are secreted from the embryo into the aleurone layer, triggering the production of enzymes.

ALKALINE STEEPING The addition of approximately 0.1% (w/v) lime (calcium hydroxide) or caustic (sodium hydroxide) or sodium carbonate (Na_2CO_3) to steep water to raise pH in order to wash out higher levels of polyphenols from barley or sorghum steeps and thereby improve beer haze (colloidal) stability. Alkaline steeping, if carefully controlled, can also be used to reduce grain surface microflora of badly contaminated grains.

ALPHA AMYLASE (α-amylase, α-1,4-glucan 4-glucanohydrolase) A principal starch-degrading (**amylolytic**) enzyme produced during malting which cleaves α-1,4 links of straight starch chains to produce glucose, maltose, maltotriose, and some maltodextrins. Cereal grain α-amylases are fairly heat-tolerant and are most active at 118–140°F (48–60°C), and pH 5.3–5.7. They can be thermostable up to 158°F (70°C).

The ability of α-amylase to hydrolyze native, raw starch granules differentiates it from other amylolytic enzymes. Alpha-amylase appears during grain maturation, declines to negligible levels in mature barley, and is synthesized *de novo* during germination in response to **gibberellic acid**. Due to the relative ease of measuring α-amylase, it is included in most analytical schemes and its activity during malting is sometimes extrapolated, often without justification, to other enzyme systems of cereals during malting.

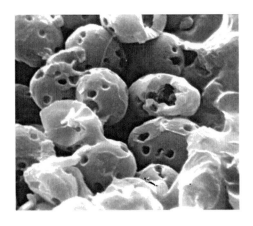

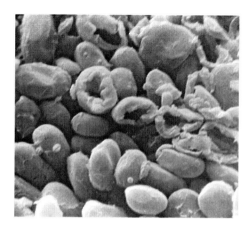

Figs. 4 (pg 8) and 5 (above): Scanning electron micrographs (SEM) of barley endosperm showing absence of small starch granules and hydrolytic pitting of large starch granules by α-amylase, indicating over-modification during malting.

ALPHA-GLUCOSIDASE (α-D-glucoside glucohydrolase, maltase, glucoinvertase) This amylolytic malt enzyme hydrolyzes α-1,4; α-1,6; α-1,2; and α-1,3 linkages of starch to liberate ß-D-glucose. The enzyme is insoluble in barley and becomes soluble during germination or after treatment with proteolytic enzymes.

AMARANTH (*Amaranthus sp.*) Amaranth is a **gluten-free** pseudo-cereal that can be malted. Malting conditions for amaranth involve steeping for 33 hours cold at 46°F (8°C) with wet (W) and air-rest (A) cycle hours of 5W–8A–8W–12A–steepout; germination for 7 days at 46°F (8°C), followed by 3 days at 59°F (15°C); kilning at 122°F (50°C) for 16 hours, 140°F (60°C) for 1 hour, and 149°F (65°C) for 5 hours.

AMBER MALT also called **mild ale** or **English pale ale malt** Amber malts are made from kiln-dried pale malts in drum roasters starting at 120°F (50°C), then increasing to 340°F (170°C) at about 5°F/minute. Amber malts are at the low end of the **roasted malt** color range—20–40 SRM (40–80 EBC) at 2–4% moisture—and are typically used at 1–3% of brewing grist in order to add brown to copper color and dry, bready, nutty, toasted, and biscuit flavors to mild ales, brown ales, porters, and stouts.

AMERICAN MALTING BARLEY ASSOCIATION, INC. (AMBA) Founded in 1938 as the Malt Research Institute, the American Malting Barley Association, Inc. (AMBA) in Milwaukee, WI is a 501(c)(3) non-profit organization charged with ensuring an adequate supply of quality malting barley for the U.S. malting, brewing, and distilling industries. Its primary focus is on the development of new varieties with improved agronomic and malting properties. The mission of AMBA is to develop six-row and two-row barley varieties broadly adapted for the barley production areas of North America with suitable agronomic, malting, and brewing performance.

AMBA provides annually about $500,000 to numerous U.S. state and federal institutions for applied barley breeding and basic malting research studies. AMBA maintains a recommended list of U.S. two-rowed and six-rowed malting barleys on their website, ambainc.org. The 2014 approved malting barley variety list includes 17 two-rowed spring barleys, 2 two-rowed winter barleys, and 8 six-rowed spring barleys.

AMBA membership is currently composed of 2 multinational breweries, 40 U.S. craft breweries, 7 multinational malting companies, 4 U.S. distilleries, and 4 North American craft malthouses.

(See Appendix B for recommended characteristics from the American Malting Barley Association)

AMERICAN SOCIETY OF BREWING CHEMISTS (ASBC) The American Society of Brewing Chemists is a 501(c)(3) non-profit organization founded in 1934 to improve and bring uniformity to the brewing industry on a technical level. ASBC was originally formed from the Analysis Committee of the United States Brewers Association (USBA) to focus primarily on malt analysis. ASBC is represented by individual and corporate members worldwide representing large and small brewers, consultants, government agencies, academics, distillers, vintners, and those working in allied industries. The web address is asbcnet.org.

AMYLOLYTIC (STARCH-DEGRADING) ENZYMES Amylolytic enzymes include **α-amylase, ß-amylase, α-glucosidase**, and **limit dextrinase**. In the malt laboratory, malt enzymes are extracted in sodium chloride or buffered solution and measured at 68°F (20°C) on finely ground malt flour. These analysis conditions are specified by the **ASBC** and others to provide a common reference method for measuring malt enzymes, but do not necessarily relate to real brewing practices, which vary widely in grind, flour dispersion, and wetting in mash water, mash temperature, thickness, and enzyme/starch contact time.

AMYLOPECTIN Amylopectin, which constitutes 70–80% of barley starch, is a branched molecule whose branches consist of α-1,6 linkages to chains of α-1,4-linked D-glucose residues. The degree of polymerization (DP) of amylopectin is of the order of 10,000 to 100,000 with an average chain length of 20–25 glucose residues.

AMYLOSE Amylose, 20–30% of barley starch, is a smaller linear molecule of 500 to 1,000 degree of polymerization (DP), with D-glucose residues joined mainly through α-D-1,4 linkages.

APPARENT EXTRACT (AE) Because alcohol is less dense than water, it artificially lowers extract measurement readings. Apparent extract is measured and then corrected to **Real extract** (RE).

ASSORTMENT also called **Grain Sizing or Percent Plumpness** Grain kernel size impacts the ratio of starch to husk, rate of water uptake in steeping, and the malt extract potential. Screens with slot widths of 4.5/64 in. (1.79 mm), 5/64 in. (1.98 mm), 5.5/64 in. (2.18 mm), 6/64 in. (2.38 mm), 6.5/64 in. (2.58 mm), and 7/64 in. (2.78mm), like those available from the **American Society of Brewing Chemists (ASBC)**, are used to measure the "plump" and "thin" assortment distribution of a representative sample of grains after harvest.

Oftentimes, cereal grains are separated by size prior to malting. In the case of two-rowed barley, kernels less than 5.5/64 in. (2.18 mm) or 6/64 in. (2.38 mm) are usually cleaned off by screening and sold as animal feed with the culms and dust. Six-rowed barleys are screened over a 5/64 in. (1.98 mm) screen.

Barley kernels <6/64 in. are called thins, and those >6/64 or 7/64 in. are called plump. Plump barley kernels can yield higher extract, lower protein, slower water uptake and may require a half or full day longer malting time for complete modification than thin kernels. For these reasons it is desirable to malt different kernel sizes separately, if possible.

AUTO-ANALYZER A continuous, air-bubble-segmented flow analyzer equipped with various heating, mixing, and detector modules that have been collaboratively tested by the **ASBC** technical committee and find use in malt laboratories for analysis of: specific gravity, α-amylase, diastatic power, ß-glucanase, pH, color, viscosity, wort ß-glucan, FAN, and soluble protein.

AWN The hair-like projection on the tip of the barley kernel of varying lengths of different varieties, which can be used to help identify varieties. Also called the barley beard, awns are cleaned off during threshing operations.

BALL ROASTER One of the earliest roasting devices developed in Germany in the 19th century as an enclosed, rotating, ball-shaped cast-iron vessel. Ball roasters have also been called "K-balls" for "kugel," the German word for ball. They have since been replaced with rotating, cylindrical **drum roasters**.

BARLEY *(Hordeum vulgare)* Barley was first cultivated in Northern Syria around 7500 bc (Hornsey, 2003) and is grown in more diverse ecologies than any other grain from beyond the Arctic Circle to the tropical plains of India, the salt banks of the Nile, the desert soils of Australia, and the high mountain valleys of the San Luis Valley of Colorado. Worldwide, barley ranks first in use for malting and fourth in total cultivated cereal crop acreage.

Malting barley acreage in North America has fallen in the past decade, by 20% in Canada and as much as 50% in the U.S.. Higher yielding varieties have certainly offset some loss of acreage but much of the reduction is due to disease, like *Fusarium* head blight, net blotch, and stripe rust in the Midwest. Adding insult to injury, a new disease, Ug99 or African stem rust, is looming on the horizon for North America barley growers. Additionally, large U.S. brewers have reduced their barley carrying inventory from the standard 18–24 months of years ago to little more than one year of inventory today.

In addition to disease and reduced inventory, alternative crops like corn and soybeans planted for biofuel production have displaced substantial acreage in North and South Dakota and Minnesota. Some recent relief has come from provisions in the 2008 Farm Bill which increase barley loan and target rates, providing a potential $137 million more to U.S. barley growers over 10 years, which is expected to increase barley's competitiveness with other crops.

Today, both two-rowed and six-rowed varieties make up the genus *H. vulgare* and are used for malting, brewing, and distilling. In 1986, total U.S. barley acreage was around 11 million acres with 22% used for malting and 51% for feed; in 2012, total barley planting was reduced to 3 million acres, of which 57% was consigned to malting and 31% to feed. Reduction in feed barley acreage is due, in large part, to increased feeding with corn and Dried Distillers Grains (DDG).

Total malting barley needed by U.S. brewers is estimated at 115 million bushels (2.76 million tons), of which craft brewers used 17.9% (20.6 million bushels) to make 5.7% of the total U.S. beer production in 2011. In 2014, the Brewers Association estimates that craft brewers will buy nearly 25% of total U.S. malt production and, as craft beer nears 10 and 20% of total U.S. beer volume, malt usage by craft brewers will be more than 30 and 50%, respectively, of total U.S. brewing malt consumption (Brewers Association, 2014).

New barley varieties are patented by breeders under the Plant Variety Protection Act of 1970, allowing them 25 years of proprietary ownership of the new variety.

Fig. 6: Photo of two-rowed and six-rowed barley heads.

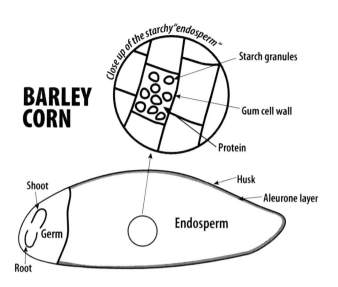

Fig. 7: Diagram of barley kernel (courtesy Institute of Brewing and Distilling).

BARLEY, TWO-ROWED *(Hordeum distichum)* A variety of barley on which only the median spikelets (at 12:00 and 6:00 on a clock face) are fertile forming seeds or grains, while all four lateral spikelets (at 2:00, 4:00, 8:00, and 10:00 on the clock face) are infertile and produce no grains. Two-rowed malting varieties generally produce larger, more uniform kernels, have higher starch content, lower nitrogen and ß-glucan, and produce higher malt extract, than six-rowed cultivars. The ideal protein range for two-rowed malting barley is 10–13%.

AMBA-recommended two-rowed spring and winter varieties in 2014:

ABI Voyager—developed by Busch Agricultural Resources, LLC in Fort Collins, CO. ABI Voyager has high yields in the intermountain region of the western U.S. with improved resistance to Spot Blotch, which may allow it to be grown further east than most North American two-rows.

AC Metcalfe—AC Metcalfe was registered in 1994 at the Agriculture Canada Experimental Station at Brandon,

Manitoba. Metcalfe is a cross of earlier Canadian varieties, Oxbow and Manley, with superior agronomic yield, lodging resistance, net blotch resistance, stem rust loose smut resistance, and tighter hull adherence than Harrington (the variety that it replaced).

CDC Copeland—Developed by Dr. Bryan Harvey at the University of Saskatchewan Crop Development Centre (CDC) in Saskatoon, Canada, Copeland has higher yield, better stem rust resistance, stronger straw strength and lower ß-glucan than Harrington, but is slightly more susceptible to loose smut and common root rot.

CDC Meredith—Developed by Dr. Brian Rossnagel at the **CDC** in Saskatoon, Canada, Meredith has similar lodging resistance, plumpness and resistance to loose smut, but is slightly later maturing and lower test weight than AC Metcalfe.

Charles—The first winter barley bred at the University of Idaho Agricultural Experiment Station and the first recommended by AMBA for adequate malting quality and brewing characteristics. In pilot scale testing by AMBA in 2000 and 2002, Charles malt was higher in extract, α-amylase, and plumpness than a reference Anhueser-Busch spring barley.

Conlon—Released by the North Dakota Agricultural Experimental Station in 1996, Conlon is early maturing with large, plump kernels, high test weight, and resistance to powdery mildew and net blotch. Conlon shows high agronomic yield, but is more prone to lodging and moderately susceptible to spot blotch. Conlon is best adapted to western North Dakota and adjacent western states.

Conrad—Released by Busch Agricultural Resources in 2005, with higher than average yields in Idaho.

Harrington—A two-rowed malting barley variety developed by the University of Saskatchewan in 1986 that yields high malt extract but is susceptible to net blotch and scald leaf diseases, stem rust, and loose smut. Harrington has very loose hulls and a tendency for kernel skinning unless carefully threshed.

Hockett—A rough-awned, two-rowed dryland malting

barley with early heading and plump kernels. While superior to Harrington and AC Metcalfe in yield, Hockett's best feature is its retention of test weight and lower grain protein percentage under drought.

Merit—Merit was developed by Busch Agricultural Resources, as a middle-maturing, medium density head barley adapted to the intermountain West. It is resistant to net blotch and moderately resistant to scald.

Merit 57—Released by Busch Agricultural Resources in 2009, Merit 57 is high yielding, late maturing, with good lodging and scald resistance, but is more susceptible to Barley Yellow Dwarf (BYD) virus. Merit 57 malt produces high levels of malt extract, α-amylase, and diastatic power, similar to Merit. It is well adapted for spring planting in the intermountain region of northern California.

Moravian 37—The original Moravian was brought to Coors Malthouse and Brewery from Czechoslovakia in 1949. Moravian 37, a descendant of that first Moravian selection, is a two-rowed spring barley variety patented and released by Coors in 2001, with good yield and test weight, early heading and short straw with resistance to lodging.

Moravian 69—Another Coors barley similar to Moravian 37 with better heat stress tolerance and faster modification in the malthouse.

Pinnacle—A variety released by NDSU in 2007, with very high agronomic yields, high test weight, early maturity, and resistance to lodging.

Scarlett—A spring barley variety developed in Germany, Scarlett has very good suitability for malting and brewing purposes. Scarlett is an early maturing variety with good lodging resistance and very low protein content.

Wintmalt—A winter two-rowed malting barley developed in Germany, with good winter hardiness.

BARLEY six-rowed *(Hordeum hexastichum)* On an ear of 6-rowed barley, three single-flowered spikelets grow on each node of the stalk or rachis. The triads at each node alternate on sides of the stalk, giving six rows around the circumference of the stalk with two median spikelets at the 12:00

and 6:00 analog clock face positions, looking down the ear, and four smaller lateral spikelets at the 2:00, 4:00, 8:00 and 10:00 positions.

The fact that 2/3 of the kernels, in the lateral positions are generally smaller than the median kernels at harvest, causes some commercial maltsters to screen and separate different sized kernels when they receive the barley, malting them separately and re-blending as necessary to meet customers' specifications.

AMBA-recommended six-rowed varieties in 2014

Celebration—A midwestern six-rowed variety, well adapted for Minnesota, North Dakota, Idaho, and Montana, with early to medium maturity, early heading, and medium-short straw height.

Innovation—Innovation was originally a feed barley, broadly adapted, that exhibits good kernel plumpness in the upper Midwest and irrigated regions of western North Dakota and eastern Montana.

Lacey—Developed at the University of Minnesota Agricultural Experimental Station in 1999, Lacey is a high-yielding barley with test weight and plumpness and average protein.

Legacy—Developed at Busch Agricultural Resources in 2000, Legacy is higher yielding than older variety Robust, well adapted to Minnesota, North Dakota, Idaho, and Montana.

Quest—Quest is the first variety released by the University of Minnesota Agricultural Experimental Station with improved resistance to *Fusarium* head blight, accumulating about half the level of deoxynivalenol (DON) as standard six-rowed varieties.

Robust—An older six-rowed variety released by the University of Minnesota in 1983, with medium height and straw strength, slowly being replaced by new varieties like Tradition and Lacey.

Stellar-ND—Released by NDSU with very high yield potential, similar to the highest yielding varieties currently grown with very strong straw and very good malting and brewing characteristics. Stellar is adapted to a wide area in the upper Midwest and Northern Great Plains.

Tradition—A Busch-Ag Resources barley variety released in 2003, Tradition is a spring-type midwestern, six-rowed malting barley well adapted for Minnesota, North Dakota, Idaho, and Montana with medium relative maturity, medium-short height, and strong straw.

BARLEY four-rowed (*Hordeum hexastichum* var. *nutans*) A variation of six-rowed barley with lax ears that bend over or "nod," causing grains to overlap each other. In 1678, Sir Robert Moray wrote,

> "Malt is there made of no other Grain but Barley, whereof there are two kinds, one which hath four rows of Grains in the Ear, the other two rows. The first is the more commonly used; but the other makes the best malt."

Bere (pronounced "bear") is a tall, high nitrogen, four-rowed barley which is still grown on the isle of Orkney in Northern Scotland, where a kind of scone, "beer bannocks," is made from it. At one time, Bere was the most important barley grown in the Highlands and Islands of Scotland, giving its name to the villages of *Bere-wic* (Berwick) and *Bere-tún* (Barton), which each mean "barley farm" or "barley village" (Hornsey, 2003). Orkney homebrewers still brew with malted bere because of its distinctive taste, said to have a pleasant smoky flavor and slightly bitter after-taste. Recently, the Arran Distillery has produced a malt whisky made entirely of this old, heritage variety, and the Valhalla, a microbrewery on Unst, one of the Shetland Isles in Scotland, brews a Shetland Ale using bere barley. Colorado Malting Company (Alamosa, Colorado) planted Bere in 2013 for multi-year seed increase to "see what this old variety looks and malts like!"

The four lateral florets of the barley head that partially fill with starch in six-rowed barley, and do not fill at all in two-rowed barley, are the only kernels that develop in four-rowed barley. These kernels are thin, misshapen and more difficult to control in malting, although they may impart unique flavors.

> "I would like to see more barley raised in our county and offer nice, clean, four rowed Scotch seed barley at $1.00 per 100 lbs. I am willing to contract for No. 1, clean barley raised from this seed at $1.00 per 100 lbs. delivered at brewery. Screenings returned.
> 24-2w. A. COORS.

Colorado Transcript (Golden), April 13, 1898.

A newspaper clipping from the *Colorado Transcript* (Golden, Colorado), April 13, 1898, reveals that Adolph Coors was appealing to local Jefferson County farmers to grow and deliver "four-rowed Scotch barley" to his 25-year old malthouse and brewery in Golden. The Coors barley contract program eventually changed to 100% two-rowed barley varieties developed and patented by Coors, and still exists as the world's foremost farmer-to-brewer barley supply-chain agreement.

BARLEY HARVEST Whereas most production barley fields are harvested mechanically with large combines (cutting, threshing, and cleaning are combined in one mobile machine), small plots are harvested manually by either direct head cutting or swathing. Ideal harvest moisture for malting barley is 10–15%.

BARLEY KERNEL **also called barley grain or corn** The barley kernel is the caryopsis, or one-seeded fruit of the barley plant containing living tissues, the **embryo** and **aleurone layer,** as well as the non-living **endosperm** where barley **starch, protein** and cell wall **ß-glucans** are deposited during kernel growth.

Fig. 8 (above): Immature barley kernels. Fig. 9 (below): Foreign seeds in barley sample.

BARLEY AND OTHER CEREAL QUALITY CHECKS PRIOR TO MALTING Barley, or other malting cereal grains, must be quality-assured at harvest, purchase, and any time prior to malting. The key performance indicators for barley quality include assessments of varietal purity, **embryo** viability, and absence of molds, **mycotoxins**, and other contaminants.

Elements of the Federal Food, Drug, and Cosmetic Act—Food Safety Law, the EPA Federal Insecticide, Fungicide and Rodenticide Act, and Good Manufacturing Practices (GMP)—can be referenced in contracts to prevent any adulteration of the purchased grain.

Grain deliveries should be identified with harvest date and origin of grain, date of delivery, truck or railcar id number, grain type, and variety name, and total weight as delivered.

Other cereal grain quality parameters might include: free of foreign odors, grain color (visual or instrumental), **Falling Number** analysis, frost damage, heat damage, foreign seeds, **pre-harvest sprout damage**, presence of mold in the seed crease, damaged and diseased kernels, skinned and broken kernels, **ergot, deoxynivalenol (DON), moisture, protein, test weight, screen assortment, Germinative Energy (GE), Germination Capacity (GC), Water Sensitivity (WS)**, and minimum/maximum grain age since harvest.

When re-sampling and analyses are called for, that should be clearly spelled out in the contract; for example, dark grains may be temporarily stored and checked in the lab for mold and germination vigor prior to purchase acceptance.

Regarding the increased risk to small producers of having to deal with sub-standard cereal quality at harvest, Wendell Banks of Michigan Malt (Shepherd, Michigan) explains:

"Even with intensive crop management, developing a strategy for diverting poor crops is essential to making craft malting sustainable. Developing brewing processes that address this challenge—i.e. step and decoction brewing—as well as distilling, represents the best opportunity for us to us to address this and make our malting operations sustainable.

The actual process of malting is but a small part of the supply chain that eventually produces beer. The amount of time, effort, and capital required to manage the supply end far exceeds the task of turning barley into malt. Starting with quality barley makes the chore easy. Maintaining a supply of quality barley, storing it, and financing it represent a significant part of the malt production chain and the cost brewers and distillers pay for their malt."

BARLEY YELLOW DWARF VIRUS (BYDV) A plant disease transmitted by aphids which can cause yellowing and dwarfism of infected plants (barley, wheat, maize, oats, and rice) severe enough to impede normal grain-filling, resulting in thin kernels, or killing the plants, wiping out entire crops. Planting late so that barley plants are still in the field during late season frosts can kill the aphids that carry BYDV. Aphid insecticides are also used to control BYDV.

BASE MALT   also called brewers pale malt   The largest portion of malt in a mash that contributes most of the active enzymes, fermentable sugars, and soluble nitrogen to the wort. Base malts contain enough starch, enzyme activity, and yeast nutrients that they can constitute 70–100% of the mash bill of materials in brewing and distilling. Most base malts are made from barley, but increasingly, other cereals are being malted to fill this role, with or without the addition of exogenous (fungal or bacterial) enzymes.

In a 2005 study, Weyerman Malting Co. made four batches of the same recipe of Pilsner beer in their 2-bbl pilot brewery, varying only four different barley varieties used to make the base malts. The barley varieties used in the test were Alexis, Barke, Scarlett, and Steffi. The base malt in all four brews made up 90% of the grist. The results of this test were that beer flavor differences were found between the four barley varieties tested.

In another study, a large U.S. maltster/brewer examined seven different growing areas of the same barley variety through malting, brewing, and beer flavor (personal communication). This study, performed over two successive crop years, found that even when barley quality parameters (variety, protein, plumpness, GE, GC, etc.) were standardized and selected between growing areas as much as possible, the beer drinkability scores for the seven areas produced repeatable differences in beer drinkability, or likeability. This study indicates that local growing conditions, or terroir, can also impact barley, base malt, and beer quality parameters.

Base malts for brewing and distilling should be selected for flavor, extract, enzymes, and fermentability.

BEER   Malt from cereal grains is primarily used in the production of beer, although beer can be made without malt, for example Guinness and Heineken beers in Nigeria. Next to water, malted cereal grains constitute the most important and expensive beer ingredients, providing all of the proteins which, in their various forms, create the aroma, flavor, mouthfeel, color, alcohol level, clarity, and foam characteristics of beer.

In addition to a refreshing and healthy beverage, beer has served as a toothpaste ingredient, antimicrobial mouthwash, enema, vaginal douche and dressing for wounds (Darby et al., 1977). Beer's role in medicine and human nutrition continues to be elucidated, namely in relation to B vitamins, minerals, and antioxidants, and protection from coronary heart disease (Bamforth, 2002).

BETA AMYLASE (ß-amylase, α-1,4-glucan maltohydrolase) A starch-degrading exo-acting **amylolytic enzyme** that cleaves non-reducing ends of **amylose** (straight) and **amylopectin** (branched) starches to produce maltose. Its name is derived from the fact that during hydrolysis there is a temporary inversion of the hydrolyzed α linkage to the ß position.

The enzyme exists in the endosperm of mature cereals in free (active) and latent (inactive) forms. The inactive form is activated and solubilized by proteolytic enzymes during germination or by agents that reduce disulfide bonds.

Beta-amylase is not dependent on gibberellic acid for development and, like α-amylase, is of secondary importance during malt modification but critical during mashing.

Beta-amylases are less heat-stable than alpha-amylases, and this property allows for the brewer to control beer fermentability, mouthfeel, and flavor by carefully controlling temperature, pH, and mash thickness.

BETA-GLUCANS (ß-D-glucans) All malting barleys contain about 3–7% ß-glucans, and providing that the malt is uniformly modified, that is, where ß-glucan levels are uniformly reduced to less than 0.4%, very few ß-glucan related problems should be encountered.

The major group of structural polysaccharides are the ß-D-glucans, which can be subdivided into two fractions, **cellulose** and **non-cellulosic ß-glucans**. Cellulose is found in husk. Barley cell walls contain ß-glucans (70–75%), **pentosans** (20–25%) and other minor components.

The influences of ß-glucan on the malting and brewing processes and quality of finished product are manifold. The ß-glucans not only determine the rate of malt modification by their presence in cell walls in an insoluble form, but their excessive presence in the finished malt can affect extract yield, wort separation, beer filtration, and beer foam and haze properties.

The rate of enzymic degradation of cell wall ß-glucan during malting is influenced by the rate and extent of water penetration in the kernel, the amount of moisture at steepout, barley variety and barley protein content.

High molecular weight ß-glucans in the mash tun can form aggregates with starch granules and protein, reducing wort separation and extract yield. The viscosity of wort is closely related to soluble ß-glucan content but may not reflect molecular properties. Addition of brewing adjuncts, such as maize or rice, which contain little or no ß-glucans, can dilute the ß-glucans from malt and improve the rates of wort and beer filtration.

There is limited information on the beneficial effects of ß-glucans in beer. It has been suggested that soluble ß-glucans may improve the non-biological stability of beer by coating and increasing the solubility of particles and play a role in foam stability and beer flavor.

APPROXIMATE BETA-GLUCAN CONTENTS OF CEREALS

CEREAL	ß-glucan, %
Barley	3–7
Brown Rice	0.1–0.2
Corn (maize)	–
Milled Rice	0.1–0.2
Millet	–
Oats	5–8
Rye	2–4
Sorghum	1–2
Triticale	1–2
Wheat	1–2

BETA GLUCANASE (**ß-glucanase**) There are many different enzymes involved in the hydrolysis of barley endosperm cell wall ß-glucan. These include: (1) endo-1,4-ß-D-glucanase also called cellulase, which hydrolyzes the endo ß-1,4-glucosidic linkages and is confined to the husk where most of the activity can be accounted for by microbial infection; and (2) endo-1,3-ß-D-glucanase, which is specific to ß-1,3 linkages. The activity of this enzyme is not dependent on gibberellic acid and may be important for initial depolymerization and viscosity reduction of soluble ß-glucan. (3) endo-1,3;1,4-ß-D-glucanase has also been referred to as laminariase, endo-barley-ß-glucanase, and mixed-link ß-glucanase. This enzyme is not found in mature barley, and develops during germination only after stimulation by gibberellic acid. Its function during malting, and continuing into mashing if kiln temperatures are low enough, is to degrade ß-glucans after they have been solubilized by endo-1,3-ß-D-glucanase. It is heat-labile and becomes quickly denatured above 149°F (65°C). There are also several exo-ß-glucosidases that hydrolyze ß-glucans to glucose.

Several commercial ß-glucanase preparations of non-malt (bacterial or fungal) origin are available for use when brewing with barley or poorly modified malts. These are used to degrade high molecular weight ß-glucans, thereby improving wort separation and beer filtration. Commercial ß-glucanase preparations that contain amylase and protease activities are more effective than "pure" enzyme extracts.

The objective of the barley breeder, grower, and practical maltster is to ensure good quality barley and malt in order to avoid potential brewhouse mash or distiller's wash problems which may require the addition of bacterial or fungal enzymes.

BETA-GLUCAN SOLUBILASE   A set of enzymes, which degrade cell wall materials that enables other ß-glucanases to digest ß-glucans directly.

BETA-GLUCOSIDASE   (**ß-D-glucoside glucohydrolase, cellobiase, laminaribiase, cellulase**)   A series of enzymes that hydrolyze the non-reducing ends of ß-glucan molecules to yield glucose. These enzymes may originate in microbial contaminants found on the external parts of grain, rather than the cereal grain itself.

BISCUIT MALT   Biscuit malt is made in a drum roaster from kiln-dried malt by heating at 250–300°F (120–150°C) for 30–45 minutes to yield final malt color of 25–50 SRM (50–100 EBC) with an orange hue and biscuit, toast, and nutty aromas and flavors when used in beer at 1–10%.

BLACK BARLEY, OR ROASTED BARLEY   Sound, uniformly sized barley at 10–16% moisture is loaded into a roaster, heated to 175°F (80°C), then internal drum temperature is increased at 2oF/min. to 300oF (150oC). The temperature is raised to 355°F (180°C) for one hour. To prevent charring of the grain surface the malt is sprayed (quenched) with water, then heated to 445°F (230°C) for 30 minutes and quenched with water again. Final roast temperatures are very close to combustible temperatures, which requires extreme caution and constant attention.

Direct heating is removed from the drum at this point and frequent sampling and visual inspection by the experienced roaster operator determines how long to wait for full color development to 300–800 SRM (600–1600 EBC) before quenching a final time with fine sprays of water (10–30 gallons/ton of grain) and cooling prior to dumping from the roaster for storage or bagging.

BLACK MALT, BLACK PATENT MALT Black malts are made from kiln-dried, or "white" malts, that have been lightly dried in a conventional flat kiln to around 6–7% moisture to allow safe storage until they are needed for roasting. The grain selected for black malt can be of inferior quality, for example short-grown malt (3 or 4 days in germination), because it is used for color and toasted flavor, not starch or other malt qualities.

After the dried malt is loaded into the drum roaster, roasting begins at 165°F (75°C), then raised to 280–300°F (140–150°C) in 60 minutes, when the grain temperature reaches 300°F (150°C) it is quenched with water sprays to prevent charring. The temperature is then raised to 340°F (170°C) for 30 minutes, followed by another water quenching, producing white steam rising from the malt. Temperature is then gradually increased to 430°F (220°C), followed by another quenching which produces steam vapors with a bluish-gray color, swelling the grains and giving them a polished surface sheen. Like black barley roasting, frequent sampling and visual inspection by the roaster determines how long to wait for full color development before the final water spray quenching. All roasted malts must be cooled to less than 85°F (30°C) before they are loaded into an analysis bin. Black malts can have colors of 400–600 SRM (800–1200 EBC) with deep red, dark brown, or black hues.

BLENDING One of the main differences between large, established malting facilities and those of newer, smaller craft malsters, in addition to size, is the former's ability to store many malts of different qualities, age them and precisely blend them to meet customer's specifications.

Malt blending allows large malt companies to blend over- and under-modified batches, crop year changes, and grains from individual areas, growers, and harvest dates that deviate from the statistical mean in some way. In addition final blended malt bins, these malsters will have many smaller analysis bins that hold finished batches long enough to perform complete malt analyses prior to loading into large blend bins.

By closely examining a handful of individual kernels from a malt sample, it may be possible to see, for example, whether the 30 SRM colored malt, as delivered, shows uniform coloring then it was probably made during a single kiln run, or if it looks more like salt and pepper, it could be a blend of dark and pale colored malts. Whereas both malts produce 30 SRM color in a warm water extract of milled malt, these sorts of differences might make a difference in the amber, red, brown, and black hues in the malt extract and resulting beer or alternative whisky.

Recent Australian research (Evans, 2012), has shown that there may be corollary benefits achieved by blending malts. Laboratory blends (40–60%) of two malts with high and low levels of various malt quality parameters showed that percent extract was linear between the high and low level malts, whereas lautering performance, as predicted by ml/25 minutes, ß-glucanase activity, and wort viscosity; and fermentability performance as predicted by α-amylase, diastatic power, and Kolbach index, may show slight synergistic improvements when two different malt qualities are blended together, presumably due to the higher enzyme activity of one of the malts.

The grain industry has also been known to blend raw grains to reduce measurable levels of contamination from **smut**, *Fusarium* **head blight, DON**, etc.

The antithesis of blending is the notion of **terroir**, or local origin, that many craft maltsters can promote, while large blending malthouses usually do not. Local sourcing has limited blending options, however, when things aren't right.

BREWERS ASSOCIATION (brewersassociation.org, Boulder, CO) The Brewers Association (BA) is an organization of brewers, for brewers and by brewers, originally organized in 1978 as the American Homebrewers Association, whose purpose is "To promote and protect American craft brewers, their beers and the community of brewing enthusiasts." More than 2,000 U.S. brewery members and 43,000 members of the American Homebrewers Association are joined by members of the allied trade, beer wholesalers, individuals, other associate members, and the Brewers Association staff to make up the Brewers Association.

Recently, the BA published a six-page white paper entitled "Malting Barley Characteristics for Craft Brewers" (2014) with the stated intent to improve malted barleys available to craft brewers with:

- distinctive flavors and aromas
- lower free amino nitrogen ("FAN")
- lower Total Protein
- lower Diastatic Power ("DP")
- lower Kolbach Index (ratio of Soluble Protein to Total Protein, or "S/T")

BREWERS SUPPORTING AGRICULTURE (BSA) A project started by the Valley Malting Company in Hadley, Massachusetts which couples local breweries and distilleries with specific fields of malting barley or wheat that is grown specifically for that brewery and malted by Valley Malting Co. In its fourth year, the 15 brewer members of the BSA pay earnest money each spring and agree to purchase about one acre of grain (~2,000 lbs.) after harvest. When needed, money is provided to growers when purchasing seed for specific varieties of interest to local brewers and distillers.

The BSA will soon add heritage landrace barley and wheat varieties to the list of sponsored varieties.

BREWING AND MALTING BARLEY RESEARCH INSTITUTE (BMBRI) Located in Winnipeg, Canada, the BMBRI was founded in 1948 to provide research grants for breeding of new malting barley varieties.

BROWN, OR PORTER, BLOWN, OR SNAPPED MALTS A traditional malt made from green malt that was used for brewing porters and "stout" porters (eventually shortened to "stout") which were the dominant UK beer styles in much of the 18th and 19th centuries. Grains were kilned in thin layers, 1–2 inches deep, on a woven wire deck heated by burning wooden poles or "faggots" made from hornbeam, oak, ash, or beechwood in a three-stage heating cycle of: 1) warm (200–250°F, 100–120°C) for 1–2 hours; 2) cooled with water sprays and manually turned; 3) roasted at 300–340°F (150–170°C) for 30 minutes which caused the kernels to burst with a loud "snapping" sound, like popcorn (Brown, 1983). Fresh-brewed porters would have been very smoky in flavor and they were often aged for many months in huge wooden vats.

Brown malts today are made from green malt roasted in a drum kiln, therefore they do not snap nor are they smoky. Single setpoint roasting temperatures of 250–280°F (120–140°C) produce brown malt colors of 50–70 SRM, 100–140 EBC.

BRUMALT, OR BRÜHMALZ Brumalts are made by increasing green malt moisture to 48% by spraying and raising germination temperature to 104–122°F (40–50°C) for the final 36 to 48 hours of germination by covering with a tarp or turning off germination fans.

After transferring to the kiln, the malt is further "stewed" at high moisture and low temperatures less than 140°F (60°C) for a few hours then slowly raised through drying temperatures (140–170°F, 60–77°C) to cure for 3–4 hours at 176–194°F (80–90°C). Brumalt has a color of 15–20 SRM (30–40 EBC), and retains much of its enzyme activities so it can be used at a high proportion of grist in dark beers.

BUCKWHEAT *(Fagopyrum esculentum)* A member of the grass, not cereal family, Buckwheat has triangular-shaped seeds that can be added as an adjunct or malted and used to make **gluten-free** beer and foods.

A malting schedule for buckwheat might include steeping at 59°F (15°C) in a wet soak (W) and air-rest (A) cycle (hours) of 3W–3A–2W–1A–1W–1A–steepout at 40–48% moisture; four days germination at 59°F (15°C); kilning for 5 hours at 104°F (40°C), 3 hours at 122°F (50°C), 3 hours at 140°F (60°C).

BUSHEL Grain was initially traded in 15th century U.K. by measuring volumes of grain contained in a cylinder of 8 inches in diameter and 18.5 inches high (Hertrich, 2013). Today, all grains are weighed first, then converted to bushel units using the U.S. Department of Agriculture (USDA) standard definitions:

| Barley—48 lbs | Barley malt—34 lbs | Corn—56 lbs |
| Rice—45 lbs | Rye—56 lbs | Wheat—60 lbs |

Worldwide, there is less standardization, however, as a bushel of barley in the U.S. and Canada = 48 lbs, Australia and New Zealand = 50 lbs, and in the U.K. and South Africa = 56 lbs.

Malted barley also suffers from similar regional disparities; U.S. and Canada = 34 lbs, Australia and New Zealand = 40 lbs, U.K. and South Africa = 42 lbs.

If cereal grains and resultant malts were measured by volume, and they're not, then the malting process might seem to create increased grain volume. Malted cereal volume is higher than the volume of the raw cereal prior to malting due to internal tissue swelling by water, enzyme action, and embryo growth.

Because **malting loss** is calculated by weight, 10–20% of starting grain weight is permanently lost in the malthouse as culms, acrospires, water, CO_2, and heat.

C

CALCOFLUOR (4-methyl, 1-7-diethylaminocoumarin) A fluorescent dye that binds to ß-glucan in barley, malt, wort, and beer. Several visual and photometric methods based on this chemistry have been developed to measure ß-glucan, including the "Carlsberg sanded slab method," in which malted grains immobilized in a resin slab are sanded down and stained with calcofluor. High amounts of or uneven distribution of fluorescence can indicate under-modification or lack of homogeneity in modification.

CANADIAN MALTING BARLEY TECHNICAL CENTRE (CMBTC) A non-profit, independent organization located in Winnipeg, Manitoba that provides technical assistance, research, and training to the malting barley, brewing, and distilling industries. Website cmbtc.com.

CAPITOL CONSTRUCTION COSTS (RELATIVE TO VESSEL COMBINATION OPTIONS) Gibbons (1989) compared construction costs for various options of large-scale (30,000 tonnes/year) U.K. maltings in £-per-tonne costs. Converting these numbers to percentage relative values where separate Steep, Germination, and Kiln vessels = 100% cost/ton:

- Steep–Germination combined (SGV) with separate kiln = 109–118% cost/ton.
- Steep separate, Germination–Kiln combined (GKV) = 104–110% cost/ton.
- Steep–Germination–Kiln combined in one vessel (SGKV) = 137–160% cost/ton.

Although it may seem counter-intuitive that single SGKV malthouses actually cost more to construct (and operate) than multiple vessel malthouses, the number of units that must be built to produce an annual malt production target increases by 2–3 times because turnaround cycles are so much slower in combined operation malthouses.

Fig. 10 (above): Seeger barley cleaning screw. Fig. 11 (below): Seeger de-stoner.

CARAMEL MALTS Caramel malts can be produced from green malt in a traditional kiln by recirculating the humid off-grain air at the start of kilning and closing the fresh air intake at 140–145°F (60–62°C) which causes a degree of "stewing" of the endosperm sugars and amino acids to form melanoidin color compounds. After stewing, final roast air-on temperatures of 175–250°F (80–120°C) produce final malt colors of 6–40 SRM (12–80 EBC).

A good "stew" period depends on the kiln being fairly air-tight, keeping humidity in the recirculated air as much as possible. The higher color caramel malts can usually only be made in kilns that hold water vapor tightly and produce sauna-like conditions during the stewing period.

Caramel malts may also be made in a **drum roaster**. **Green malt** is loaded into the drum roaster which is sealed to prevent moisture escape. External heat is applied to raise internal drum temperature to 140–160°F (60–71°C) for 30–60 minutes to stew the grain. The sealed drum is then opened to exhaust the moisture and the grain is dried with warm dry air, ramping up to roast temperatures of 240–320°F (115–160°C) until target colors of 10–120 SRM (20–240 EBC) are achieved.

Caramel malts are used at 3–15% in amber colored ales, altbiers, Belgian abbey ales, red ales, cream stouts, bock beirs, Scottish ales, Vienna lagers, and light lagers, providing greater sweet caramel flavors than more highly roasted dark malts, and improving head retention and flavor stability in beer.

Caramel malts have many different names worldwide, including crystal, Cara, Carastan, Extra Special, Special B, Carapils, Carafoam, Caramunich, Caravienne, and others.

CELLULOSE Cellulose is the most abundant structural and cell wall polysaccharide in the plant world. It accounts for more than 50% of the total organic carbon in the world. On complete acid hydrolysis, cellulose yields only D-glucose linked together by ß-1,4-D-glucan linkages.

The linear structure of cellulose gives it its strength and tendency to form microfibrils, making it insolubile in water.

Most of the early studies of barley endosperm cell walls, were based on the assumption that they were composed, like most other plant cell walls, of cellulose. This view was disproven by MacLeod and Napier (1959) who found that most of the cellulose in raw barley grain (4% by weight) can be accounted for by the husk, embryo, and aleurone of the kernel, therefore the endosperm cell walls do not contain significant amounts of cellulose.

Due to the insoluble nature of cellulose, it persists largely unchanged into the spent grains. These "high-fiber" spent grains are composed of 9–10% cellulose, 19% pentosans, 10% lipids, and 30% protein and are marketed as ruminant feed.

Recent attention drawn to the importance of fiber in the diet has resulted in some brewery spent grains being used in baking breads, chips, and pet food.

As part of the Food Safety Modernization Act (FSMA), the Federal Drug Administration (FDA) is in the process of accepting comments on proposed new rules regarding feeding animals with brewery and distillery spent grains. The proposed rules, as currently written, would require that spent grain for animal feed be dried and prepackaged onsite in a manner that does not touch human hands, a ridiculously restrictive change to current methods involving wet grains delivered in bulk.

CHIT, OR CHITTING, OR "CHICK," ALSO CALLED THE ROOT The first appearance of the white root sheath or coleorhiza at the grain embryo during steeping which later produces roots in germination. The number of kernels in a 100-kernel sample showing evidence of rootlet growth at steep-out (recorded as percentage chit) is a necessary and simple measure of steeping effectiveness and grain viability.

Fig. 12: Stages of barley rootlet growth from chitting (left) to fully grown (right).

CHIT OR SHORT-GROWN MALTS Chit malts are germinated for a short time, 1–2 days, and then kilned at low temperatures to preserve enzymes. Historically, chit malts have been used to increase mash **α-amylase** levels, typically in Germany, by brewers wishing to comply with the strict ban on non-malted ingredients of the **Reihheitsgebot**.

An editorial in the Nov-Dec issue of the Journal of the Institute of Brewing seems to condemn the use of chit malts by German brewers during the First World War thusly:

> W. Windisch (1917) states that in the case of the 1915 barleys the grist used successfully consisted of 20–40 per cent, of imperfectly modified corns, equivalent, therefore, to the so-called "chit" malt, which consists of barley steeped and couched, and as soon as sprouting has commenced loaded on kiln, whilst the remainder was short-grown malt. This latter is malt germinated for a considerably shorter period than a normal one. Naturally in neither of these forms of grain has modification proceeded very far, but the Author states that he has corrected this by the adoption of certain methods in the brewery. We would point out that in pre-war days "chit"malt was condemned by the majority of German brewers as being only raw barley.

CHOCOLATE MALT Chocolate malts are made in drum roasters from kiln-dried malts (also called "white" malts), in the same manner as **black malts**, except that final roast temperatures are 10°F cooler, around 420°F (215°C) and no quenching sprays are used. Chocolate malts are roasted to colors of 250–500 SRM (500–1000 EBC) with red and brown hues and are used in dark ales and lagers, porters, stouts, and alternative whiskeys.

An experienced brewer's "secret" is that chocolate malt, by itself, will not produce much in the way of "chocolaty" flavors, but, add a little vanilla bean (the second most expensive spice in the world), vanilla extract, or artificial vanillin flavoring (inexpensively extracted from wood lignin) and taste the difference!

CLEANING GRAINS PRIOR TO STEEPING Contaminants in grain such as field straw, wood, nails, wire, stones, foreign and immature seeds, and broken kernels should be removed prior to malting. Machines available for grain cleaning include magnets, aspirators, vibrating screens, destoners, deawners, and dust cyclones.

Grain washing prior to steep-in separates small and foreign seeds, removes dust, and reduces microflora on the grain surface. One common grain washer, in the form of a steeping screw, also called an Archimedes screw, is inclined in a trough with counter-current flows of water (from top to bottom of the inclined trough) and grain (loaded into the bottom of the trough and augered up to the top), before washed grain spills over the top of the trough into an outlet chute for transport to the steep tank.

COLD WATER EXTRACT A method, principally used in the U.K., to assess malt quality by extracting a sample of milled malt in 68°F (20°C) water. While not useful for determining overall malt quality, it can be used by maltsters to select kilned malt runs that show higher cold water extract levels, indicative of elevated sugars and amino acids, and might, therefore, yield higher color when dry-roasted.

COLOR, MALT EXTRACT OR WORT COLOR The color of laboratory extracts of finished malt samples can be expressed in °L (degrees Lovibond), or the Standard Reference Method (SRM) units of the American Society of Brewing Chemists or European Brewing Congress (EBC) units. The Lovibond scale was developed in the late 19th century by Joseph Lovibond in England, and uses glass reference discs in a Lovibond tintometer. SRM and EBC units are modern spectrophotometric measures of malt and beer color at 430 nm wavelength. SRM and EBC methods differ from each other because they specify different sample cell path lengths (SRM = ½ inch and EBC = 1 cm). Mathematical conversions of °L and EBC units to SRM units are as follows:

SRM = (°L − 0.76) × 1.3546   SRM = EBC × 0.508
°L = (SRM / 1.3546) × 0.76   EBC = SRM × 1.97

For malt colors on the low, pale malt end of the scale, 1–10°L, the SRM and Lovibond units are roughly equal, to within one unit of each other, so they can be used interchangeably. For higher color malts, SRM color units are 10–30% higher than Lovibond for the same malt. However, it is generally agreed by most U.S. maltsters and brewers that Lovibond and SRM units are close enough to each other, and therefore are used interchangeably. Even though malt colors are often given in °L, Lovibond units, they are always measured using the ASBC SRM, never measured with an outdated Lovibond tintometer. In this text, both SRM and EBC color units are given when discussing malt extract colors.

In addition to malt extract color determined in a laboratory, many factors can determine final color at point of use, including malt blend makeup, water chemistry, wort boiling, and beer filtration. Two different malts may have the same extract color on analysis but produce visual differences in terms of red, yellow, orange or brown color, hue, and intensity, which can be assessed only by tristimulus color analysis or actual use and final product comparisons.

COLOR, QUICK ETHANOL EXTRACT COLOR METHOD Like a book, you can't tell a malt by its cover (external color). Only by milling and extracting in warm 113–158°F (45–70°C) water for more than 2 hours, in the ASBC method, can one get an accurate, collaboratively tested, measure of malt extract color. However, by finely milling a 5 g malt sample in a small electric coffee grinder, stirring for 1 minute at room temperature in 35 ml of 70% (v/v) ethanol, pouring through filter paper into a ½ inch diameter test tube and visually comparing to a malt extract color chart, like those found on several malting company websites, an approximation of warm water extract color can be made. (Ethanol concentration and extraction time may vary by malt type and color range.) Alternatively, color can measured at a wavelength of 430 nm on a spectrophotometer. This quick method has been used for malt colors between 2 and 40 SRM and must be routinely calibrated against a standard malt extract color method.

COLOR, QUICK MICROWAVE COLOR METHOD Recently, Li and Maurice (ASBC, 2013) developed a microwave-based method for quick color measurement of malts with colors between 1.50 and 3.50 SRM (3–7 EBC). The method extracts 22 g of coffee grinder-milled malt in 150 ml of distilled H2O, which is heated in a microwave at 20–40% of maximum power for 6–8 minutes. After filtering, color is measured at a wavelength of 430 nm on a spectrophotometer. This quick method is dependent on individual microwave power characteristics and must be routinely calibrated against a standard malt extract color method.

Fig. 13 (above): Marty Matruzzo of FarmHouse Malt Newark Valley, NY. Fig. 14 (below): (l-r) Eric Owens, Bill Owens, and Jason Cody in germinate-kiln room at Colorado Malting Co. Alamosa, CO.

COMMON CONVERSIONS USEFUL TO MALTSTERS
FLOW...

1 barrel/hour (bph) = 0.52 gal/minute (gpm)	1 gpm = 1.92 bph
1 foot^3/minute (cfm) = 62.43 lbs water/min	1 lb water/min = 0.016 cfm
1 foot^3/sec (cfs) = 448.831 gpm	1 gpm = 0.0022 cfs
1 gpm = 1,440 gal/day (gpd)	1 gpd = 0.00069 gpm
1 gpm = 8.0208 feet^3/hour (cfh)	1 cfh = 0.1247 gpm

PRESSURE...

1 bar = 14.7 lbs per square inch (psi)	1 psi = 0.068 bar
1 inch water gauge (wg) = 0.0361 psi	1 psi = 27.0 inch wg
1 ft wg = 0.4335 psi	1 psi = 2.307 ft wg
1 lbs/square foot (psf) = 0.006944 psi	1 psi = 144 psf

WEIGHT

1 inch^3 of water = 0.036 lbs	1 lb of water = 27.68 inch^3
1 foot^3 of water = 62.4 lbs	1 lb of water = 0.01602 feet^3
1 grain per gallon (gpg) = 17.118 ppm	1 ppm = 0.0584 gpg
1 gal water (50°F) = 8.3453 lbs.	1 lb of water = 0.1198 gal
1 ounce (oz) = 437.5 grains = 28.35 grams (g)	1 grain = 0.0023 oz = 1.0368 g

CONVEYOR AND ELEVATOR TYPES, ADVANTAGES (A) AND DISADVANTAGES (D)

Belt conveyors—A: horizontal and inclined position, low grain damage, long life, minimal maintenance, low power requirements, incline up to 20° from horizontal, can be enclosed and fitted with anti-spill rails. D: dust generated at feed and discharge points.

Screw conveyors—A: horizontal and inclined position, range of throughputs determined by flight size, pitch, and RPM. D: high power requirements, regular maintenance of bearings essential, will not handle >40% loading of trough, maximum inclination is 12° from horizontal.

Chain and flight drag conveyors—A: horizontal and inclined position, low grain damage when operated at maximum capacity, low capital cost, low power requirement, chain speed optimum <1 fps for friable malts. D: High noise levels, grain damage if run at reduced capacity, regular maintenance of chain and flights, maximum inclination 20°, requires rotation sensors to detect chain breakage

Dilute phase pneumatic conveyors—A: horizontal and

vertical positions, can transport materials through tortuous routes, material and dust fully contained, can be rigged for pressure or vacuum, although vacuum systems operate at lower velocity and larger pipe diameters. D: limited conveying capacity, high power requirements for compressed air, some damage based on roughness of pipe and number of direction changes, high capital cost, regular maintenance on air seals.

Dense phase pneumatic conveyors—A: horizontal and vertical positions, less damage than dilute phase. D: High power requirements for compressed air, high capital cost, limit in conveying capacity, regular maintenance on air seals.

Bucket elevator, high-speed—A: vertical position, enclosed material and dust, run at approximately 9 fps. D: material fed into down-leg of elevator requires higher power, requires rotation sensors to detect belt slippage or breakage, regular maintenance required to adjust belt tension and inspect buckets.

Bucket elevator, slow-speed—A: same as high-speed, material is fed into upleg side of elevator which runs approximately 5 fps, lower power requirements, D: same as high-speed.

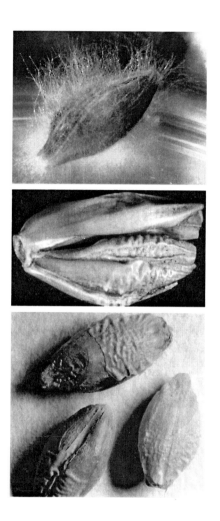

Fig. 15 (above): Barley with mold mycelium growth in petri dish. Fig. 16 (center): Barley kernel showing field chemical damage. Fig. 17 (below): Barley kernels skinned during harvest.

CONTRACTS FOR GRAIN SUPPLY Annual contracts to supply malting grains after harvest, can include, but are not limited to:

- Full names and contact information for grower,

- transportation agents, and maltster.
- Grain type, variety, contracted amount, contract period, fields of origin, estimated harvest dates, restrictions on blending, specific sites, and timeframes for delivery.
- Location and certification of truck, bag, or laboratory scales for all weights.
- Details of seed purchase arrangements from maltster or third party.
- Open option to purchase more grain than allotted by current contract.
- Exact pricing including discounts, premiums, storage, and transportation fees.
- Written guarantee that grain is free of all liens and encumbrances.
- Reference to application of all appropriate regulatory rules, i.e. the National Grain and Feed Association (NCFA) and Food, Drug and Cosmetic Act of 1938, and Good Manufacturing Practices, as amended.
- Exact analytical methods used, i.e. **American Society of Brewing Chemists (ASBC)**.
- Specific conditions when grain will be temporarily stored, probed, and re-analyzed prior to acceptance or rejection.
- Maximum allowable levels of **mycotoxins**, aflatoxins, and **deoxynivalenol (DON)**.
- Maximum **%moisture** (and minimum if grain is artificially dried by grower).
- Grain screening distribution, **%plumpness, test weight**.
- Maximum and minimum levels of **protein**, % dry basis.
- Maximum percentage skinned and broken kernels.
- No visible mold growth, **sprout**-damaged kernels, weed seeds, or straw.
- Free from foreign or musty odors.
- Maximum percentage of foreign, damaged, and diseased and **ergot** kernels.
- Maximum percentage frost, heat, and chemical-dam-

aged kernels.
- Minimum grain **germination% (GE, GC, and WS)**.

The Brewers Association (2014)
"…believes that all U.S. malting barley stakeholders will benefit from increased custom contracting by breweries of all sizes. Contracting will provide surety to growers and maltsters as they undertake the important process of increasing the number of malting barley varieties available to U.S. brewers. Because barley has rapidly become de-commoditized in a few short decades, custom contracting with an increased number of breweries for an increased number of varieties will be an essential solution to continued U.S. barley malt market innovation and competitiveness."

COUCH, OR COUCHING The practice of heaping freshly-steeped grains into a pile in traditional **floor malting** in order to allow the grains to naturally heat up and accelerate **chitting** and germination. Following a period of couching, the grain is manually spread on the germination floor.

As described by Sir Robert Moray in 1678,
"…when the barley is sufficiently steeped, take it out of the trough, and lay it on heaps [the couch], and so let the water drain from it; then in two or three hours turn it over with a shovel, and lay it in a new heap about 20 or 24 inches deep; and this they call the coming heap, and in the right managing of this heap lies the greatest skill; and in this heap it will lie 40 hours, more or fewer, according to the fore-mentioned qualities of the grain, &c. before it come to the right temper of malt; whilst it lies in this heap, it is to be carefully looked to after the first 15 or 16 hours; for about that time the grain will begin to put forth the root, which when they have equally and fully done, the malt must within an hour after be turned over with a shovel, other-

wise the grains will begin to put forth the blade or spire also, which by all means must be prevented; for hereby the malt will be utterly spoiled, both as to its pleasantness of taste and its strength…"

Wahl and Henius (1908) described the "Warm Sweat Method" of couching during floor malting as, "The temperature of the heaps is allowed to rise high, viz., 77 to 86°F (25–30°C). The radicals develop rapidly, the acrospire very unevenly. Germination is rapidly completed."

The "Cold Sweat Method" was described as,
"The temperature of the heaps is kept low, about 63.5°F (18°C). The acrospire develops gradually and more uniformly, but germination takes longer. The cooler a heap is kept, the better is the quality of the malt regarding solubility, diastatic power and aroma. "Sweat" is the moisture which will appear on the surface of the barleycorns during germination, the vapors passing from the interior of a warm heap and condensing near the surface. The appearance of this "sweat" is a sign of healthy growth."

Couching can be performed in **pneumatic malthouses** by simply turning off CO_2 and heat evacuation fans in the steep for several hours prior to steep-out.

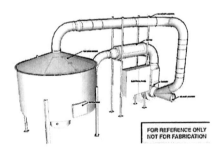

Fig. 18: Awn Engineering 1-ton malting equipment (courtesy Christian Stanley).

CRAFT MALTING EQUIPMENT MANUFACTURERS Recently, there are several designs of new craft malthouses being offered commercially: (S = steeping, G = germination, K = Kilning, V = vessel).

- **Awn Engineering & Equipment**, Hadley, MA: 1-ton cylindroconical SGKV batch system and 1-ton floor malting kiln. Email christian@valleymalt.com, Christian Stanley (413) 349-9099.
- **Blacklands Malt**, Austin, TX: 2-ton Saladin system with separate cylindroconical steeping and GKV. Website blacklandsmalt.com, Brandon Ade (530) 289-6258.
- **BrauKon CombiMalt**, 2-ton combined steeping, germination, kilning, and roasting (SGKRV) vertical cylindrical vessel, automatic controls. Webstite braukon.de/en/contact/, +49 (0) 171 62 77 584.
- **Buhler Inc. Pargem Container Malting units**, Plymouth, MN: 10-ton batch system made from forty foot shipping containers, Email mark.larson@buhlergroup.com, Mark Larson (763) 847-9900.
- **Colorado Malting Company**, Alamosa, CO: 5-ton batch SGKV Saladin box system. Email coloradomalting@hotmail.com, Jason Cody 719-580-5084.
- **Kaspar Schulz Brauereimaschinenfabrik & Apparatebauanstalt e.K.**, Bamberg, Germany: 2, 5, 10 & 25-ton batch cylindroconical steep and separate drum GKV configuration. Email binkert@kaspar-schulz.de, Jörg Binkert +49 (0) 951 60 99 22.
- **MacDonald Steel Limited**, Cambridge, Ontario, Canada: 200-kg pilot malting to 1-ton Saladin box systems. Website macdonaldsteel.com, John Cressman (519) 620-0400 ext. 3219.
- **Malterie Frontenac Inc.**, 5-ton SGKV Saladin system. Website www.malteriefrontenac.com, Bruno Vachon (418) 338-9563.
- **New York Craft Malt LLC**, Batavia, NY: 1, 2, 3 & 4-ton cylindroconical SGKV batch and separate steeping systems, imported. Email ted@nycraftmalt.

com, Ted Hawley (585) 813-5389.
- **Western Feedstock Technologies,** Bozeman, MT: 100-kg suspended-rail malting system. Email blake@montana.edu, Tom Blake (406) 599-4889.

Fig. 19: Kaspar Schulz cylindroconical steep tank and rotary Germinate-Kiln vessel (GKV) (courtesy Kaspar Shulz).

Fig. 20: Buhler Pargem 5-ton SGKV malting unit in 40 ft shipping container (courtesy Buhler Inc. Plymouth, Minnesota).

CRAFT MALTSTERS GUILD (NORTH-AMERICAN-MALTSTERS-GUILD@GOOGLEGROUPS.COM) Formed as an online discussion group in 2012, the Craft Maltsters Guild was formed by Andrea Stanley of Valley Malt (Hadley, MA):

"The mission of the Craft Maltsters Guild is to promote and sustain the tradition of craft malting in North America, provide services and resources to the Association's members, and uphold the highest quality and safety standards for Craft Malt."

There are currently more than 60 members in the Craft Maltsters Guild.

The Craft Maltsters Guild website, www.craftmalting.com, defines "craft malt" as:

>…a finished malt product, produced from a variety of grains including but not limited to barley, wheat, rye, millet, oats, corn, spelt, and triticale. Craft Malt is in particular made using a majority (greater than 50% by weight) of locally grown grains as inputs, meaning grains grown within the region of the Craft Malthouse, are used to produce Craft Malt therewith. Craft Malt is made without the use of gibberellic acid (GA) or other chemical additives during processing.

Fig. 21: Craft Maltsters Guild logo.

A Craft Malthouse is defined as a business producing Craft Malt, with upper limitations of 10,000 metric tons and lower limitations of 5 metric tons of combined product sold per year. A Craft Malthouse must maintain proper legal standing with its country, state, and local governments including but not limited to: business registration and licensing, taxation, safety, and other regulations pertaining to the production of grain, malt, and foodstuff, as appropriate. Malthouses are not considered Craft Malthouses if ownership by other non-Craft Malthouses exceeds 24%."

CRYSTAL MALTS To produce crystal malts, fully germinated green malt at 44–48% moisture is loaded into a drum kiln, external heating applied, and rotated at 122°F (50°C) briefly to dry excess moisture from the grain surface. The drum is then sealed to prevent evaporation and the internal temperature is slowly increased to 142–162°F (61–72°C) during which the malt enzymes are at work "stewing" the starches to sugars and liquefying the interior of the kernel. By manual sampling and squeezing individual kernels between the thumb and index finger, the maltsters determines whether enough liquefaction has occurred prior to going to curing temperatures.

At this point, exhaust vents are opened and internal drum temperatures are raised to curing temperatures of 250–350°F (120–177°C), depending on desired malt color. When the color is reached, the malt is cooled rapidly and the sugars crystallize in unfermentable colored **dextrins**.

Colors of crystal malts are sometimes signified in product names in Lovibond units, for example Crystal 10L, 40L, 120L, up to 200L.

When drum roasters are not used, caramel or crystal malts can be made in the traditional manner by loading green malt into a conventional flat kiln in a thick, even layer, wetting it down, covering with a tarp to slow evaporation and heating to 140–160°F (60–70°C) for 30 minutes to two hours to facilitate endosperm liquefaction. After ensuring that the endosperm is completely liquefied, the tarp is removed and roasting temperature raised to 212–320°F (100–150°C), depending on desired color.

Some kilns (e.g. steam-heated) are capable of maintaining higher levels of humidity, or allow for direct steam injection, facilitating the stewing process.

Crystal malts add malty, toffee, biscuity, nutty, and toasted flavors and can improve head retention, flavor stability, and add amber, brown, or red color hues to beer.

Fig. 22: Liquefied endosperm after stewing in crystal malt process (photo courtesy of Roger Putman).

CULMS, ROOTLETS, OR COOMBES The dried rootlets (3–5% of total barley weight) attached to finished, kilned malt that must be removed during malt cleaning, deculming, or dressing. Some undesirable compounds, such as sulfur dioxide, nitrosamines, and phenols can accumulate in high levels in culms, up to 100–200 times that found in cleaned malt, stressing the importance of efficient malt cleaning. In the past, culms were sometimes distributed over the top of malt during storage as an absorbent, super-dry sponge to slow or prevent uptake of atmospheric water by the malt below. Culms, and other malt byproducts, are sold as high protein (25–30%) animal feed.

The importance of properly cleaning or dressing malt in historic England was discussed by White (1860):

"Prior to the time of Henry VIII, no priest was allowed to eat or drink at any place appointed for the sale of ale, called ceapealethetum. The manufacture of malt in the olden time was so much cared for that by an old act of Parliament, any maker of malt who knowingly sold malt not properly separated from the roots or dross, was punished with fine and the pillory."

CURING The final step in malt **kilning**, when the grain is heated at specified grain moisture, air temperature, and humidity to develop flavor and color.

CYTOLITIC (CELL WALL DEGRADING) ENZYMES Brown and Morris (1890) recognized the importance of cytolytic enzymes to cell wall breakdown in germinating cereal, and Hopkins and Krause (1937) had postulated that a "number of cytases" are involved in cell wall changes. However, these malting scientists assumed that the cell walls were made of cellulose and described cell wall degrading enzymes as "cellulases" or "cyto-hydrolysts."

Later, these enzymes were classified as **pentosanases** and **ß-glucanases**.

DEOXYNIVALENOL, ALSO CALLED DON, OR VOMITOXIN DON is a mycotoxin produced in cereal grain crops when infected with **Fusarium head blight (FHB)**, the same microorganism that can cause **gushing** in beer. DON has caused illness (hence the name "vomitoxin") and even death in animal feed and human DON-contaminated bread outbreaks in the past.

Mold growth and DON production can increase during malting and DON concentrates in aqueous processes, therefore it persists into the final beer rather than spent grains.

Some maltsters and brewers will not accept any grain or malt with detectable levels of DON (detection limit 100 ppb by GC/MS). Others set maximum DON levels in grain up to 2,000 ppb. In the European Union (EU), maximum DON levels have been established as 1,250 ppb for barley and malt, 750 ppb for cereal flour and pasta, and 200 ppb for processed infant foods.

In the U.S., the Food and Drug Administration (FDA) has only set "advisory levels" of 1,000 ppb in final food products. It is believed that the FDA shied away from setting absolute limits of DON in foods, because the Environmental Protection Agency (EPA) had recently been severely excoriated and sued by the U.S. apple industry for their mishandling of the infamous carcinogen Alar (Daminozide) plant growth regulator, causing Uniroyal to withdraw that chemical from the market in 1989.

Serious epidemics of FHB and DON occurred in the Midwestern U.S. resulting in virtually none of the barley being used for malting from the 1993, 1994, and 1995 crop years.

In studies done at NDSU, forced air on-farm storage of Fusarium infested barley at 75°F (24°C) was effective in reducing FHB infection levels of barley (Beattie et al., 1998). DON levels were also lower in malt produced from barley stored under these conditions.

As stated by Andrea Stanley of Valley Malt,

"The challenges of operating a micro-malthouse are substantial. Malting equipment is not easy to procure and the learning curve to manipulate grain into malt is steep. For all the challenges we face, the largest is finding a good reliable source of quality grains. As many of us know, "Good malt starts in the field." For many of us, finding the correct varieties to grow in our region is a huge hurdle. In New England we only consider growing barley and wheat varieties that have Fusarium resistance. It is the #1 reason why we reject an otherwise suitable lot of grain. We have seen DON levels over 8 ppm and many times these numbers discourage farmers from ever trying to grow grains again.

With all of the positive goodwill going into these emerging grain and malt industries, we cannot forget that all of it could be dampened out by the threat of DON. No matter how high the demand and how great the premium a crop may bring, if you are going to lose it 3 years out of 5 to DON, you are not going to continue grow it and take that risk. Corn is a much better bet. If funding is not put into researching resistant varieties for growing DON free grains, this renaissance will never get off the ground."

DEXTRINS "Dextrin" or "dextrine" was a term first used in the U.K. in 1830 to describe a gummy component of starch, later modified to unfermentable sugars from malt. Polymers of glucose with more than three glucose units linked by α-1,4 and α-1,6 linkages and are not normally fermentable by brewing yeasts, resulting in increased mouthfeel and fullness of beer. In normal brewer's worts, about 75% of the wort sugars are fermentable, leaving 25% of wort sugars as non-fermentable dextrins. In light beer worts and most distillery mash preparations, nearly 100% of the malt and adjunct sugars are fermentable, leaving practically zero dextrins.

Distilling yeasts can also ferment maltotetraose (four linked glucose units), which is normally present at about 10% in wort.

DEXTRIN MALT **Caramel malt** that is stewed at low temperatures (less than 120°F, 50°C) to produce non-fermentable dextrins for mouthfeel and improved beer head retention with no dark color or enzymic activity. Final moisture is generally higher (6–8%) for dextrin malts along with very low color, 1.5–2.0 SRM (3–4 EBC).

DIASTASE A collective term for all the starch-degrading enzymes in malt first coined by French scientist Anselme Payen in 1833. The word diastase comes from the Greek word diastemia, meaning "I separate," referring to the ability to separate starches into sugars. The diastase group includes **α- and ß-amylases, limit dextrinases**, and **α-glucosidases**. Diastatic enzyme activity during malting is limited by time and moisture levels in order to retain and release as much activity as possible into the mash.

DIASTATIC POWER, DP A measure of the total diastase or starch hydrolyzing power of malt. Barley variety notwithstanding, final malt DP is highly, positively correlated with total barley protein, whereas kiln temperatures and malt color are negatively correlated to final malt DP. In the U.S. and U.K., DP is reported in degrees Lintner (°Lintner) while in Europe, units are from the Windisch-Kolbach method (°W-K). The relationship between the two methods is:

DP (°Lintner) = (°W − K + 16)/3.5

Maltose equivalent (ME), another, though rarely used unit of DP activity, is calculated as four times °Lintner.

Ishida (2002) compared the diastatic power ratio of several malted cereals to malted barley and found that wheat produces about 50% of the DP of malted barley, sorghum 25%, finger millet 15%, and barnyard, pearl, foxtail, and proso millets as well as rice, produce less than 10% of the total diastatic power that barley can produce when malted.

The Brewers Association (2014) established a DP maximum of 150° Lintner, rationalizing that

"… current barley varieties have sufficient enzyme potential and higher levels are not helpful for craft brewing. In fact, some craft brewers advocate for the development of malting barley varieties with lower DP levels,"

while carefully guarding against under-attenuation in fermenting.

DIMETHYL SULFIDE (DMS) A powerful aromatic compound that imparts a sweet creamed-corn smell to lager mashes. In finished beer it imparts a malty quality or, at higher levels, the taste of cooked corn or black olives. Higher protein barleys and cooler kiln roast temperatures may lead to higher malt DMS levels. Beer DMS levels of 30–45 ppb can contribute mouthfeel and sweetness, but levels higher than 50 ppb can produce dominant and objectionable flavors. Final concentration of DMS in beer is determined by barley variety, kilning conditions, whirlpool and wort cooling temperatures and times, and fermentation conditions.

DISC MILL A mill used for grinding grain, metals or chemicals that has two round discs with grooved or serrated surfaces facing each other. At least one of the discs rotates to grind material fed between them. Small disc mills are used in most grain quality laboratories to provide finely ground flour for analysis. Larger units are used in corn and peanut butter wet-milling operations and have recently been tested in combined milling–mashing processes in brewing and distilling research labs.

DISTILLERIES WITH FLOOR MALTINGS IN SCOTLAND (list assembled by Andrea and Christian Stanley of Valley Malt, Hadley, MA)

- Balvanie Distillery, Dufftown, www.thebalvenie.com
- BenRiach Distillery, Newbridge, www.benriachdistillery.co.uk
- Bowmore Distillery, Isle of Islay, www.bowmore.com
- Highland Park Distillery, Isle of Orkney, www.highlandpark.co.uk
- Kilchoman Distillery, Isle of Islay, www.kilchomandistillery.com
- Laphroaig Distillery, Isle of Islay, www.laphroaig.com
- Springbank Distillery, Campbeltown, www.springbankwhisky.com

A typical Scottish distiller's floor malting process:

Steep for two days with 12 hour soak, 12 hour air-rest, 12 hour soak, 12 hour air-rest. Steep-out moisture 40–45%. Laid on the growing floor for five to six days where it is turned and raked regularly. Taken to the kiln when the moisture content is 30–38%, peated for 10 hours to 20 ppm phenols and dried to 5% moisture in 35 hours. The finished malt is transferred to a bin for storage prior to use.

DOCKAGE Dockage is defined in the ASBC Methods of Analysis (14th edition) as:

"All matter other than barley that can be removed from the original sample by use of an approved device according to procedures prescribed in Federal Grain Inspection Service (FGIS) instructions. Also, under-developed, shriveled, and small pieces of barley kernels removed in properly separating the material other than barley and that cannot be recovered by properly rescreening or recleaning."

DOORNKAAT STEEPING SYSTEM A patented 20th century system of steeping that injected compressed air at the base of vertically installed tubes in the steep tank. The tubes were fitted on rotating arms in the tank. Airflow through the tubes swept barley along with it, thereby continuously **rousing** the grain and dissolving oxygen into the water.

DORMANCY The tendency of all plant seeds to protect themselves from premature germination before, during, and immediately after ripening and harvest. Variety, inherent levels of the plant hormone **abscisic acid** (ABA), agricultural practice, time to grain ripening, grain surface microflora, and weather all play roles in a determining a seed's degree of dormancy. Cool and damp growing seasons can lead to higher levels of grain dormancy, as can warm temperatures 12 to 16 days after barley ear emergence, whereas warm temperatures later during plant development actually decreases dormancy.

Dormancy is assessed with lab methods for **germinative energy (GE), germinative capacity (GC)**, and **water sensitivity (WS)** and can range from none (with a risk of pre-germination in the head, windrow, or storage bin) to profound dormancy requiring additional post-harvest storage time. Contact with attemperated air, or extraordinarily, via applications of chemical agents like hydrogen peroxide, ethanol, salicylate, fusicoccin, sulphuric acid, n-caproic acid, or sodium azide can break dormancy.

A variety of U.K. barley introduced in the 1990s, Triumph, was known to show higher than average dormancy periods, up to six months, compared to other barleys, four to six weeks, under similar conditions. Storing 12–13% moisture Triumph barley at 80–85°F (27–30°C) for up to two weeks then cooling to <40°F (5°C) with fans at 4.8 cfm per ton of barley (0.15 m3/min per tonne) helped to shorten the dormancy period for this variety and others (Palmer and Taylor, 1980).

DRUM MALTING A type of **pneumatic malting** equipment first developed in 19th century France by Nicholas Galland in which germination is carried out in a rotating drum with automatic filling and discharge. The advantages for drum malting are that no manual or other mechanical means are needed to turn the grain and grain filling and discharge is facilitated by the drum angle and supporting platform. Possible disadvantages reported for drums used as steep or germination vessels is that throughput capacity can be limited, compared to other forms of pneumatic malthouse. In **roasted malt** production, only the kilning is done in drums.

Fig. 23: Paddles and central quenching pipe inside drum roaster at Joe White Maltings Australia. (Photo courtesy of Roger Putman.)

DRUM ROASTER A rotating cylindrical drum that replaced the direct wood-fired flat-kiln and 19th century **Ball roaster** technologies for making a wide range of dark malts from freshly germinated green malt or kiln-dried "white" malt. Roast times and temperatures up to nearly 500°F (260°C) are best controlled by an experienced roaster operator that can make roast completion decisions quickly by hand-sampling and inspecting grains every few minutes during the roast cycle. Malthouse drum roasters are often made from converted coffee roasters or 55-gallon drums with paddles inside to ensure grain mixing and prevent grain sticking to the internal drum surface and charring. Spray water for quenching can be introduced quickly through a central spindle in the drum.

Fig. 24: Home-made gas-fired drum roaster at Colorado Malting Co., Alamosa, CO.

DRYING BARLEY In wet climates, or during cool, wet harvests, grain moisture at harvest may be high enough to cause mold growth during storage and the grain's latent dormancy levels increase to protect the growing embryo from premature germination in the field. Long-term storage of wet grains (>12%) can reduce their germinative capacity (GC), germinative energy (GE), and make the grains unsuitable for malting (Palmer, 1989).

Barley harvested at 18% moisture or higher should be dried immediately; 16–18% moisture barley can be stored a short time with ventilation before drying; 14–16% moisture a little longer time with continuous ventilation; 12–14% moisture barley can be kept a long time with ventilation; and barley with <12% moisture may be safely stored a year or more in ventilated or unventilated bins or flat grain storage (Bamforth and Barclay, 1983).

Dormancy periods of up to 6 months in duration have been observed in wet crop conditions. In the mid-1960s, Scottish maltsters discovered that artificial drying barley immediately after harvest, using malthouse kilns at low temperatures 100–115°F (40–45°C) for 12–14 hours reduced seed dormancy. Continuous cross-flow vertical grain driers at higher temperatures 140–150°F (60–65°C) can be used, but it has been shown that barley dried in deep-bed, low-temperature kilns, recovers faster from dormancy than that dried in continuous driers (Bathgate, 1982).

Care must be taken when artificially drying grain by ensuring that a free and complete exhaust of moisture out of the drier occurs. If there are any dead-air spots while heating, then grain **dormancy** may actually increase.

Factors that determine safe drying of grain include 1) seed variety; 2) grain maturity (immature grain being extremely sensitive to heat damage); 3) initial grain moisture (grain at 16% moisture may tolerate drying air temperatures as high as 165°F [75°C] in a vertical drier, whereas 25% moisture grain should not see drying air temperatures above 138°F [59°C]); and 4) grain drier construction.

DUMAS NITROGEN METHOD A laboratory method for quantifying total protein (nitrogen) in cereal grains, malts, adjuncts, and worts by total combustion in the presence of oxygen and measurement by thermal conductivity detector.

DUST EXPLOSIONS Grain dust can be as dangerous as coal dust or sawdust when four key conditions are met: 1) the dust is combustible, like grain dust; 2) dust is suspended in air at a high concentration; 3) there is an oxidant, like atmospheric oxygen; 4) there is an ignition source, such as fire, electrostatic discharge, electrical arcing, or hot surfaces like overheated bearings.

The second condition, suspension of dust in the air can cause a secondary, more dangerous explosion of dust that was suspended in air by the shock of a first, much smaller explosion. The risk of a secondary, more massive shock-mediated explosion is the main reason to practice good hygiene in any grain-handling facility and clean all dust spills as soon as they happen.

E

EINKORN WHEAT *(Triticum monococcum)* The German word for "one corn" given to a variety of hulled wheat which is similar to **emmer** and **spelt**. Einkorn can be malted and is primarily grown in the mountainous regions of Europe and North Africa. Some Belgian brewers add einkorn for increased color, head retention, and vanilla and honey flavor notes. Faro is the term for baking flour produced from einkorn, spelt, and emmer.

Einkorn is difficult to malt because the embryos are easily damaged. Malted einkorn can give beer increased head retention and mild flavors of vanilla.

EMBRYO The living tissue of cereal grains, constituting about 5% by weight, that contains the root system, scutellum, and acrospire. The embryo produces endogenous gibberellins that serve as signals to other plant tissues, like the aleurone, to produce and secrete enzymes into the endosperm starch and protein reserves. Improper grain threshing during harvest or hot, still storage conditions have been known to damage embryos, thereby upsetting normal rootlet and acrospire growth and malt modification.

From the maltster's viewpoint, the most important function of the hydrated embryo during steeping and germination is the production and secretion of endogenous gibberellic acid, which will activate the aleurone layer to produce a mixture of hydrolytic enzymes necessary for endosperm modification (Palmer, 1973).

The processes of modification are initiated by the embryo during the first 2 to 4 days, depending on germination vigor, and thereafter, the presence of the embryo is not required for malting to proceed (Kirsop and Pollock, 1958).

EMMER WHEAT *(Tricum dicoccoides)* A hulled variety of wheat grown in Albania, Austria, Greece, Italy, Spain, and Turkey. Emmer is very difficult to mill, as the husk tightly adheres to the grain.

Mayer et al., 2011 malted three varieties of Italian emmer and one **einkorn** in micro-malting trials using short immersion steeps (2–4 hours first 59°F [15°C] immersion, 20–22 hour air-rest, followed by a 2 hour second immersion). Steep-out moisture was 47%. Germination (59°F, 15°C) was approximately 96 hours, and kilning was 15 hours at 131°F (55°C), 4.5 hours at 162°F (72°C), and 3.5 hours at 180°F (82°C). Final malt and brewing qualities were found to be acceptable.

ENDOSPERM The largest tissue of cereal grains, representing 75% of the total weight of barley. The endosperm contains the food reserves of starch and protein, protected by cell walls rich in **ß-glucans**. Maltsters generally desire to solubilize nearly 100% of the cell walls, nearly 50% of the protein matrix, and little (around 10%) of the starch granules contained in the endosperm during malting.

ENGINEERING STANDARDS FOR BARLEY AND MALT
- Barley = 45–52 ft3/ton = 38–44 lb/ft3
- Steeped Barley = 65–75 ft3/ton = 26–30 lb/ft3
- Steep immersion aeration airflow (rousing) = 1–3 cfm/ton
- Steeping CO_2 evacuation during air rests = 10–25 cfm/ton
- Green Malt = 73–85 ft3/ton = 23–27 lb/ft3
- Floor malting bed loading = 5–10 lb/ft2
- Pneumatic germination compartment bed loading = 25–120 lb/ft2
- Germination airflow = 200–500 cfm/ton
- Heat generated from barley respiration in germination = 200,000 kcal/ton
- Ideal germination air temperature differential across bed = 5–10°F (3–5°C).
- Water addition by sprays during germination = 10–

25 gal/ton
- Kiln bed loading = 20–100 lb/ft2
- Kiln airflow = 1,500–3,500 cfm/ton until breakthrough then 700–1,500 cfm/ton
- Dried Malt = 55–65 ft3/ton = 30–36 lb/ft3
- Angle of repose of barley = 23° from horizontal
- Angle of repose of malted barley = 26° from horizontal
- Silo hopper valley angle >35° or two hopper bottoms of 45° from horizontal

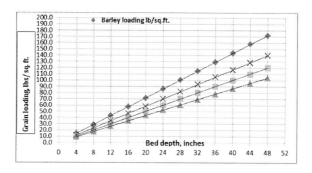

Fig. 25: Bed loading (lb/ft2) of barley, steeped barley, green barley malt and dried barley malt for various bed depths from 4 to 48 inches.

ENZYMES All enzymes are proteins, but not all proteins are enzymes. Enzymes act as catalysts by increasing the rate of specific metabolic reactions hundreds of times faster than they would occur in the absence of enzymes.

After sufficient hydration of the endosperm, the enzymes of cereal grains play a primary role in malt modification. Enzymes hydrolyze and mobilize cereal seed reserves to supply nutrients to the growing embryo and release them into the hot water mash. Proper kilning and grain milling ensures a level of residual enzyme activity after kilning which continues to catalyze the release of carbohydrate and protein into the aqueous extract during mashing.

Overall malting and mashing enzyme activity levels are affected most readily by barley variety, protein level, and kilning conditions. Germination conditions can also play a role in enzyme development rate and final **green malt** enzyme activity levels. Higher germination temperatures, 62–68°F (17–20°C), develop malt enzymes more rapidly, but may suppress maximum levels that are achievable at lower temperatures 53–62°F (12–16°C).

Control of grain moisture and temperature during kilning, as well as mash thickness, pH, and temperature in the brewhouse or distillery, are critical process variables to maintain the enzyme's structural integrity and activity. The loss of structural integrity and conformation is known as denaturation, which permanently destroys an enzyme's ability to stay in solution and "grab" onto enzyme-specific substrates, like starches, ß-glucans, and proteins. Wort boiling in brewing, denatures all malt enzymes, causing them to settle as "trub" in the kettle. Because malt and adjunct extracts are not usually boiled in distilleries, malt enzymes can remain active into fermentation.

As the renowned British malting chemist H.E. Armstrong pointed out in 1890: "Enzymic action or enzymosis… appears always to involve decomposition by means of water…" hence the term "enzymic or enzymatic hydrolysis."

The carbohydrases important to barley malting can be classified into two groups: starch degrading (amylotic) and cell wall degrading (cytolytic) enzymes.

ERGOT *(Claviceps purpurea)* A disease of barley, wheat and rye that can produce alkaloids that are toxic to humans and animals. Cold spring weather followed by damp, rainy summers can produce the characteristic dark purple coloring that can infect a few grains or replace the entire head of grain with a fungal sclerotia, making it highly visible and easy to control.

ETHYL CARBAMATE A carcinogen formed during distillation from the precursor, methyl glycoside, found in some malted barleys. Efforts are underway by distilling barley breeders to select varieties that produce low or no methyl glycoside during plant growth.

EXTRACT The collective term for all of the materials—**sugars, dextrins, proteins, ß-glucans** and **amino acids**—that are extracted into warm water from malted cereals. Malt analyses in laboratories yields the malt's potential extract in agreed-upon standard methods and are reported as "%Extract, as is" (total extract by weight including malt moisture) and "%Extract, dry basis" (total extract by weight with malt moisture removed mathematically). Actual extract, measured in the brewhouse, includes **Original Extract** (OE), **Real Extract** (RE), and **Apparent Extract** (AE).

F

FALLING-NUMBER TEST A laboratory method used to assess whether grain has sprouted (pre-germinated) prior to malting. **Alpha-amylase**, one of the earliest enzymes to be developed during germination, reduces the viscosity of a suspension of milled grain flour via starch hydrolysis. This activity is measured by the time it takes for a plunger or ball to fall through the prepared suspension. The faster the fall, the higher the alpha-amylase activity and degree of pre-harvest or grain storage sprouting, which can make the grain unsuitable for malting.

FAN LAWS that apply to steeping, germination and kiln-fan design and operation:
- Air volume delivered is proportional to fan speed. Volume = Constant × Fan Speed
- Air pressure developed is proportional to the square of fan speed. Pressure = Constant × Fan Speed^2
- Fan power required is proportional to Volume × Pressure. Power = Constant × Fan Speed^3
- Volume of air delivered by a fan reduces as resistance to flow increases
- Fan efficiency rises and then falls as flow increases, peaking at 80% for a centrifugal fan

- Power requirement rises then levels out as flow increases
- Choose fan to operate near, but not at, peak efficiency
- The larger the fan and slower impeller speed = greater percent fan efficiency
- Axial flow fan is used for low pressure applications (0–78 inches of water, 0–2.8 psig)
- Centrifugal fan is used for high pressure applications (78–780 inches of water, 2.8–28.4 psig)
- Twice the air flow or bed depth increases fan power requirements eightfold
- Ten percent higher airflow or bed depth requires thirty percent more power
- Ductwork Pressure drop = Constant × Ductwork Resistance × Velocity^2
- Green malt bed pressure drop is approximately 0.5 psig for every inch of bed depth, depending on type and size of kernels and compaction/loosening caused by rootlet and acrospire growth
- Target CO_2 evacuation during steep air-rests at 4.0–5.0 ft^3/min/ton for steeps at 200–300 lb/ft^2 loading
- Target aeration rousing during steep soaks at 2.0–2.2 ft^3/min/ton
- Target 250–400 ft^3/min/ton humidified air through germination bed to remove metabolic heat and CO_2 and maintain oxygen levels
- Target 1,200–2,000 ft^3/min/ton (barley weight at steep-in) of dry air through kiln bed loaded at 50–100 lbs/ft^2

FATTY ACIDS More is known about the fatty acids and subsequent flavor impacts for barley than the other malting cereals. Linoleic acid (18:2) and others in barley can be enzymically and chemically converted to form staling aldehydes with low flavor thresholds such as hexanal and trans-2-hexanal. These compounds can cause papery, cardboard, winey, and rotten fruit off-notes in aged beer.

FERMENTABLE EXTRACT, OR MALT FERMENTABILITY The proportion of total fermentable material in **wort** or malt extracts that is potentially fermentable when an excess of yeast is added and standardized conditions and times are followed. There are a plethora of fermentable extract methods that are mostly used as reference methods when comparing one malt to another.

FERULIC ACID A simple phenolic compound that is generated in malting, from malt pentosans during mashing and by hot-surface "cook-on" in brewhouse or distillery operations. Some bacteria and wild and domestic yeasts contain an enzyme that converts ferulic acid into 4-vinyl guaiacol. This pungent compound has the aroma and flavor of cloves or antiseptic considered a flavor defect in most beers, but is positively associated with the flavors of certain Belgian wheat or "wit" beers.

FLAKED CEREALS Unmalted cereal grains (barley, wheat, oats, maize, rice) that that are cooked in hot air at 400–500°F (204–260°C) and rolled into flat flakes and are added as adjuncts in brewing to impart foam stability, flavor, and mouthfeel to stouts, porters, and other beer styles.

FLAVONOIDS **Polyphenols** extracted from malt—including anthocyanins, chalcones, flavonols, and flavanols—that contribute to beer flavor, color, and haze, and act as antioxidants.

FLAVOR EVALUATION While chewing whole kernel malts can help the maltster assess kernel-to-kernel homogeneity and modification, it is generally regarded as an unreliable method of malt aroma and flavor evaluation. Typical ASBC laboratory extracts (worts) can be tasted, but they are fairly low in specific gravity and can taste watery and thin.

Voight et al. gave a presentation at the 2013 annual meeting of the Master Brewers Association of the Americas (MBAA) in Austin, TX titled "A New Approach to Malt Flavor Characterization." The authors described using 20 difference reference aromas and flavors, including vanilla, oak, coffee flowers, banana, apple, smoke, etc.

Routine flavor evaluation has been done on whole kernel malts, fine-grind flour and Congress worts, as well as malt "teas" produced by stirring 5 g of milled malt in 100 ml of hot water for a few minutes, filtering, and tasting.

According to the Brewers Association white paper (2014),

> "Today's suite of U.S.-grown barley varieties and the resulting malts has been characterized by craft brewers as being flavor-neutral and/or lacking distinctiveness. A number of Brewers Association member craft brewers have indicated that as a result of the recent varietal progression from Klages => Harrington => Metcalf, the sensory profiles of their flagship brands have evolved over time, drifting towards lower overall flavor impression and/or complexity."

The Brewers Association further states that there are currently at least three separate research efforts in the U.S. to study and understand the origins of flavor in barley and malt at Colorado State University (Ft. Collins, CO), Oregon State University (Corvallis, OR) and the USDA ARS National Small Grains Collection (Aberdeen, ID).

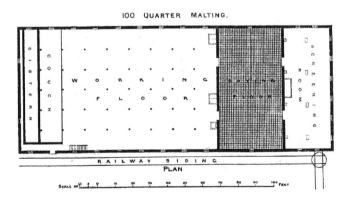

Fig. 26: 100 Quarter (44,800 lb/year) Floor malting layout (from Scamell and Colyer, 1880)

FLOOR MALTING The "lowest tech" and oldest form of home and commercial malting, dating back to the 3rd century a.d. In some parts of Africa and Asia, outdoor floor maltings with no temperature controls at all, can still be found. Floor maltings typically had a wooden or cast-iron cistern with a solid floor for single, cold-water immersion steeping for 48–96 hours with no air addition or CO_2 removal. Water for steeping was typically drawn out of a nearby river or reservoir at 40–50°F (5–10°C) resulting in very low **chitting**, or germinated kernels, at steep-out.

After steeping, the wet grains are shoveled, tipped, or wheelbarrowed into piles, or couches, of one to three feet high on the germination or growing floor. Allowing the grain to naturally warm from heat of respiration during "**couching**" facilitates water uptake uniformity and rootlet formation. After the maltster is satisfied with grain moisture and chitting levels, the couches are then broken down onto the growing floor and grain is leveled at three to six inches depth (200–400 square feet per ton of barley).

For barley floor malting planning purposes, multiplying the grain bed depth (inches) times floor area (square feet) should roughly equal 1,200. The factor, in metric equivalents, is roughly 2.8 for bed depth (meters) times floor area (meters). Other grains may require different floor malting depths and surface areas, based on botanical growth and respiratory heat characteristics of the growing grain.

In the growth, or **germination** phase, which typically takes five to seven days at 57–64°F (14–18°C), the malt is manually turned with wooden malt shovels or heavy wrought-iron rakes or plowed with electrically powered turning machines 1–2 times/day in cold weather, or 3–5 times/day in warmer weather.

White (1860) describes floor malting taking 10–16 days at 48-49°F (9°C) outside temperatures with 51–56°F (11–13°C) grain temperatures.

> "The floors are examined several times during the day. When the malt is made for the greatest quantity of saccharine matter, to suit small brewers, the rootlets should not at any time exceed three-

eighths of an inch in length, and grow slowly on the floor. In some places even during winter, where slow growth is desirable, the grain is not allowed to be thicker than one inch, and is either ploughed or turned about once in two hours. In such instances, the grain is not thrown on the kiln until the sixteenth day after emptying from cistern, when the acrospire extends nearly the whole length of the grain. Malt made upon this principle is the most valuable, both for richness of flavour and saccharine extract. If however, the malt be for pale ale brewers, or for "measure," the acrospire is forced by spontaneous heat, when during cold weather, it seldom lies thinner than four inches on the floor, and is fit for the kiln on the tenth day."

Floors may be constructed of simple materials like stone, painted brick, slate, epoxied or waterproofed concrete, or more elaborate materials, like stainless steel or the special Bavarian quarry tiles historically preferred by many early German malthouses. The very earliest floors were simply made of beaten earth or wood, with their understandable hygiene problems.

Floor maltings generally are not equipped with air conditioning units, relying instead on seasonal temperatures and the manual opening and closing of windows to regulate day and night-time temperatures of the malt on the floor, up to a maximum of about 57°F (14°C) for floor malted barley. For this reason, floor malting is not usually attempted during the hottest days of summer, though heating can be achieved by underfloor wiring or hot water radiators.

This lack of forced air movement through the grain bed causes a localized blanket of CO_2 around every kernel of grain, which slows respiration and modification and gives the malt a rich aroma and flavor that, some say, is difficult to duplicate in a modern, pneumatic malthouse.

Fig. 27 (above): Hillrock Estate Distillery (Ancram, NY) floor malting. Fig. 28 (below): Floor maltsters and their tools (Stephens, 1993), from left to right: turning fork, turning shovel, skip, bushel measure and strickle, tin shovel, and turning fork.

Fig. 29-31 (left to right): Electric turner, manual plough and power shovel at Tuckers Malting U.K. (photos courtesy of Roger Putman.)

RULES for Floor working and Kiln

1st DAY, Loading.
First plough previous days' sprinkled pieces.
Load Kilns.
Turn old pieces.
Put out young pieces as required.
Turnout old pieces again if ordered, leave ploughed always this day.

AFTERNOON.
Turn old pieces and leave ploughed.
Fork up Kiln.

2nd DAY.
Plough all required first, fork Kiln.
Turn all on Floor, put out and turn young pieces as required.
Screen steeping off Kiln or previously if Malt Order is on while Breakfast time.
Turn or plough old pieces as directed and screen.
10 a.m. fork up Kiln, fill up time Malt.

AFTERNOON.
Turn old piece, put out young one, turn and plough both.
Fork up Kiln.

3rd DAY.
Plough both pieces, turn Kiln with shovel.
Turn all on floors, empty cistern.
Plough if ordered, turn Kiln 10 a.m., fill up time Malt orders.

AFTERNOON.
Turn both pieces, leave ploughed, dress Kiln and turn if dry, if not in Morning.
Shut up tops to half required distance, same at bottom.
The Kilns are in charge of night fireman from 8.30 p.m.

4th DAY. Hotting or Curing Day.
Divide Men, some to turn hot Kiln, others to put on sprinkling water and turn the sprinkled pieces as ordered, the Men who turn the hot Kiln to fork the old pieces forward, then all join at couching.
The Kilns are turned at 8 if necessary, and at 10.30 always, heats 200 to 210.

AFTERNOON
Turn sprinkled pieces and plough them after getting the Kilns off, these pieces are ploughed again at 7 p.m. on sprinkling night only.

5th DAY same as first, and in rotation while ordered to change.

The Floors and Kilns are in charge of the Men named for each Kiln or Malting. and he is responsible to the management for any ommissions of duty on the part of the Men under him.

The times of steep and let off will be given to each Kiln Foreman, and he must adhere to the time stated, until ordered to alter.

Canvas boots should not be worn while loading the Kiln or turning grain in a growing state, as the band soles crush and damage so many more corns than the bare feet do.

These Rules subject to alteration at the discretion of management.

By order THORPE & SONS, Newark-on-Trent

Fig. 32: Floor malting rules from Thorpe and Sons, Newark-on-Trent circa 1880.

FOOD RESERVES The food reserves located in the endosperm of mature cereal grains can be divided into two groups: those required immediately for embryonic respiration and growth, and those that are stored in the endosperm in a largely insoluble form for later usage by the embryo, brewer, or distiller.

The respiratory substrates consist of sucrose and raffinose (which together constitute 20% of the dry weight of the embryo), lipids, and amino acids. These substrates are concentrated in the **embryo** and **aleurone layer** and are used during the first 24 hours of growth (MacLeod, 1969). They provide the immediate energy needed by the embryo to begin respiration and are used for enzyme synthesis in the aleurone in response to embryo-secreted **gibberellic acid** (Palmer, 1969).

Food reserves in the endosperm include **starch, protein, ß-glucan**, and **pentosan**. The chemical characteristics and physical relationship of ß-glucans, proteins, and pentosans to starch, make enzymic hydrolysis imperative prior to the use of most cereal grains in the brewing or distilling process.

FREE AMINO NITROGEN (FAN), ALSO CALLED α-AMINO NITROGEN Compounds of varying molecular weights, including ammonia, amino acids, polypeptides, and proteins, that contain a free amino nitrogen group measured by several analytical techniques. A minimum threshold level of FAN (165–175 mg/L for brewer's malt and 100–150 mg/L for distiller's malt) is critical to provide nutrients for complete fermentation, depending on wort gravity, inclusion of adjuncts, yeast source and health. FAN reaction with ninhydrin (1,2,3-indanetrione monohydrate) to produce a deep purple color is the most common method used today with FAN content recorded in parts per million (ppm) units.

Lekkas et al. (2014) found that 88% of the total yeast-utilizable nitrogen present in wort was a product of endosperm degradation during malting of 28 varieties of two-rowed barley examined. The remaining 12% was formed by proteases during mashing. They also found that two important amino acids, methionine and aspartic acid, were found to be at threshold levels for yeast nutrition.

The Brewers Association white paper on Malting Barley

Characteristics for Craft Brewers (2014) set desired malted barley FAN levels at < 150 ppm saying that

> "… development and acceptance of malting barley varieties which include lower FAN levels are important for the continued growth of all-malt beer brands. Lower FAN malts will improve product stability and promote continued geographic growth of individual all-malt beer brands. While FAN levels will, to some degree, also decrease with lower total protein and lower enzyme levels, Brewers association believes that FAN levels should be considered as an important characteristic during varietal breeding and development."

FRESH MALT, ALSO CALLED FIERY MALT As soon as malt is removed from the kiln it is called fresh or fiery malt. Not unlike coffee beans, some roasted malts may benefit from brewing use soon after kilning to retain some volatile aromas and flavors. On the other hand, pale or brewer's malt that depend on starch granule release during milling, uniform mash hydration, and enzymic hydrolysis, require a minimum period of dry, cool **malt aging**, thought to be two to four weeks, to avoid potential milling, lautering, fermentation, haze, and flavor problems.

These same problems can occur again if the malt is aged too long, or in humid, warm conditions, becoming **slack**.

FRIABILITY The simplest and least expensive measure of malt friability is achieved, with practice (and ignoring any dental concerns!) by the maltster or brewer counting out 100 malt kernels at random, and nibbling them with incisors, one at a time, from embryo to distal end of each kernel. As each kernel is chewed a mental note is made of the friability (crushability or mealiness) and sweetness of the kernel, as well as the transition point of malt modification from soft (embryo) end to hard (distal) end.

The Friabilimeter is a simple but elegant device made by Pfeuffer to measure the friability and homogeneity of finished malt, both measures of malt modification.

Friability (%) is calculated as the percent of malt sample that is crushed by the Friabilimeter in 8 minutes. Homogeneity (%) is defined as 100% minus the "All-glassy" and "Half-glassy" fractions retained (not crushed) by the Friabilimeter.

Since the Friabilimeter was introduced by Chapon and co-workers in 1979, it has found extensive use in malt quality measurement and prediction of brewhouse performance. Statistical correlations of malt friability with malt extract yield, extract viscosity, Kolbach index, beer fermentation and filtration performance, and other malt modification metrics have been well documented.

The Friabilimeter crushes a 50 g sample of dried malt in 8 minutes via a rubber roller held against a stainless steel mesh drum by a spring. Percent all-glassy and half-glassy kernels and friable flour are weighed separately and reported for each malt sample.

In addition, percent dead barley, hard ends, and case-hardened malt can be uniquely identified using the Friabilimeter (Thomas, 1986). Dead barley is defined as those kernels of barley that have gone through the steeping, germination, and kilning processes, but clearly show no signs of embryo or acrospire growth during malting. These kernels can cause brewhouse or distillery mash separation difficulties when present in very small quantities (1–2% of total kernels by weight).

Hard ends are defined as the distal ends of barley malt representing the lateral extent of modification during malting that are retained on a 2.2 mm screen.

Case-hardened malt kernels (a term first coined by Thomas, 1986) are whole or partial pieces of malt separated by the Friabilimeter which show signs of normal botanical growth as evidenced by a groove along the dorsal surface of the kernel formed by growth of the acrospire under the husk layer during germination. Case-hardened malts have been shown to be well-modified by several classical and modern methods of malt analysis, but have not been crushed by the Friabilimeter due to an interplay of malt amino acids, sugars, and kilning conditions (i.e. localized caramelization), requiring careful visual examination of all retained Friabilimeter malt fractions.

Giarratano and Thomas (1986) found, in an analysis of 69 production two-row malt samples by Friabilimeter and Calcofluor-staining methods that the Friabilimeter method was highly correlated to total protein (r = –0.64) and malt extract (r = 0.60) but was not correlated to soluble protein (r = 0.09) in malt. The Calcofluor method, however, was not statistically correlated with total protein, (r = –0.15), was correlated with soluble protein data (r = 0.34), and was somewhat less correlated to percent malt extract (r = 0.36). These differences indicate that the Friabilimeter and Calcofluor-stain methods can yield rapid and accurate information about general malt modification, but can result in different information regarding malt protein content and solubilization.

Wentz et al. (2004) tested the statistical correlation of Friability and Calcofluor staining on six-rowed barleys of malting (var. Morex) and feed (var. Steptoe) qualities. They found strong correlations of these two methods with malt ß-glucan content (extent of cell wall hydrolysis), but no correlation to protein modification.

Both the ASBC and EBC recommend checking the Friabilimeter frequently against high (80–85%), medium (70–75%) and low (60–65%) friability malt samples.

The Pfeuffer Friabilimeter sells in the U.S. for about $7,000.00 (Profamo Inc, Rancho Palos Verdes, CA.)

***Fusarium* HEAD BLIGHT, OR SCAB** A common fungal disease in cereals, also called scab, that colors grains white to pink to purple, can produce **mycotoxins**, like deoxynivalenol (DON) or vomitoxin, and can also cause **gushing** in beers, if not controlled. *Fusarium* and other fungal organisms can also cause premature yeast flocculation (PYF) in beer.

The U.S. Wheat and Barley Scab Initiative (USWBSI), founded in 1998 at Michigan State University (East Lansing, MI), seeks to develop effective control of *Fusarium* head blight (scab) and production of mycotoxins in wheat and barley. More than two dozen U.S. universities and the

USDA-ARS cooperate in a wide range of research projects aimed at achieving this goal. Funding for the USWBSI in 2014 was very nearly eliminated, but has since be re-allocated $4.6 million in federal funds.

GECMEN MALTING SYSTEM An early vertical malting system composed of a series of metal plates and spindles that rotated into closed or open positions and allowed grain to drop into other tanks below. Turners were not needed because the grain was stirred by dropping from tank to tank.

GELATINIZATION The swelling of individual malt and solid adjunct starch granules at characteristic temperatures, the gelatinization temperature, that must occur prior to liquefaction and saccharification by hydrolytic malt enzymes.

Gelatinization temperature ranges of cereals in a 4% aqueous slurry as measured by loss of birefringence, from Shelton and Lee (2000).

CEREAL	GELATINIZATION TEMPERATURES
Barley	132–142°F (56–61°C)
Brown Rice	154–163°F (68–73°C)
Corn (maize)	149–163°F (65–73°C)
Milled Rice	156–167°F (69–75°C)
Millet	145–154°F (63–68°C)
Oats	127–138°F (53–59°C)
Rye	120–133°F (49–56°C)
Sorghum	154–165°F (68–74°C)
Triticale	138–145°F (59–63°C)
Wheat	129–138°F (54–59°C)

Fig. 33: Michigan Malt Shepherd, MI.

Fig. 34: Simple but effective craft malting process control at Colorado Malting Company, Alamosa, CO.

GERMINATION Germination has been defined, in a strict botanical sense, as the process that starts with water imbibition, proceeds through intermediate phases of enzyme activation and mitosis, and ends with elongation of the radicle or rootlet (Wellington, 1966). The germinating seeds uptake of water and oxygen and production of heat and CO_2 occurs by the same mechanism whether the seed is in soil or in the malthouse.

The term "germination" in the malting process however, generally covers the growth phase that occurs after steeping and before kiln drying. Paradoxically, the radicle emergence, or "chitting," which signifies the end of botanic germination, usually occurs prior to the end of steeping and before malting germination has begun. "Germination" as defined in the malting context, rather than the strict botanical definition, is used throughout this text.

Critical control points in the germination compartment of a malthouse include properly humidified and attemperated air to remove evolved CO_2 and heat while preventing drying of grain for the entire four to seven day germination cycle.

Because the germination process step is the longest of the three separate malting cycles, including steep-out, **couching** (if done), germination time, green malt transfer to kiln, sweeping grain from corners, floors and equipment surfaces (often called "mucking out"), and compartment cleaning, depending on grain type and modification rate, the turnaround rate of germination vessels largely determines the annual capacity of a malthouse.

Germination air humidification is accomplished by drawing fresh incoming air through water spray chambers and mixing this, via air diverters and shutters, with recirculated off-grain air depending on ambient air temperature, relative humidity, and germination phase.

To prevent matting of grain rootlets during germination, and to provide fresh air to the respiring embryo, the grain must be manually or mechanically turned (e.g. with helical screw turning machines) every 6–12 hours during germination.

It is most crucial that when grain moisture target level of 40–48% is reached (depending on grain type and size), moisture levels must be maintained as high as possible throughout germination until full modification is achieved.

Adequate spray humidification chambers or adding water by spraying during germination will help reduce moisture loss during germination. Once the **acrospire** growth and kernel "rub-out" indicate that modification is complete, then grain can be **withered** by turning off sprays prior to kiln loading.

At the end of germination, green malt kernels should have rootlets that are 1–2 times kernel length, acrospire growth ideally ending at ¾ to 1 kernel length and not broken through the husk, moisture levels are generally 1–2% lower than at steep-out and the green malt should smell fresh and green, not sour, moldy, or musty. If grains were not desiccated at this point in time, the growing embryo, rootlets and acrospire would continue to consume all of the food reserves from the endosperm to produce new, green, photosynthesizing plant tissues. **Kilning**, the next malting process step, provides the means for cessation of plant germination by desiccation, stabilization of developed enzymes, and malt flavor development.

Hygiene in steeping vessels, germination compartment, and spray chambers are critical control areas in malthouses. Loose grains must be removed, steep tanks cleaned with 1.2% NaOH caustic at 145°F (63°C), and mold growth controlled in steeping and germination by periodic drying or applications of sterilizers like sodium hypochlorite (bleach) with active chlorine content of 250 ppm or higher, through water pressure nozzles, called "pop guns," or by manual application and scrubbing. Thorough rinsing is required following cleaning treatment with chemicals to prevent flavor taints in malt.

Fig. 35: Raking malt at Warminster Floor Malting U.K.

Fig. 36: Floor malt kiln with turners at Crisp Malting U.K. (Photos courtesy of Andrea Stanley, Valley Malt, Hadley, MA.)

GERMINATIVE ENERGY (GE), GERMINATIVE CAPACITY (GC), AND WATER SENSITIVITY (WS)

The **ASBC** Methods of Analysis (14th edition) Barley-3 details methods for the ability of barley to germinate as measured by (A) Germinative Energy (GE), which measures the vigor of germination at the time the test is taken; (B) Germinative Capacity (GC)—Hydrogen Peroxide (H_2O_2) Method (International Method), in which the grain's overall potential to germinate in 72 hours, (percent living kernels), is measured by the use of a dilute solution of H_2O_2 to overcome latent grain **dormancy** in freshly harvested grain; and (C) Germinative

Energy, Germinative Capacity, and Water Sensitivity—Simultaneous Determination, which in addition to measuring GE and GC, measures Water Sensitivy (WS), a sample's (100 kernels) ability to germinate in the presence of both a minimal (4 ml) and excessive (8 ml) amount of water on a filter paper in a petri dish.

The dishes are covered and kept at room temperature, 64–70°F (18–21°C) for 72 hours. Kernels are examined and counted for rootlet growth (chitting) every 24 hours, and if chitted, are removed from the dish. After 72 hours, the total number of chitted kernels from the 4 ml water dish is recorded as % Germinative Energy (GE) and the number of unchitted kernels in the 8 ml water dish recorded as % Water Sensitivity (WS). If WS is greater than 20%, then the grain is considered water sensitive and the duration and cycle of water soaks and air rests in steeping must be adjusted in the malthouse.

These methods should be carried out prior to committing barley to production steeping, because steep soak and air rest times may need to be adjusted to obtain even and complete germination. Cereal type, variety, growing conditions, and time post-harvest can affect results, so familiarity with specific malting grains is paramount to interpreting results and comparison to previous crops.

Thomas et al. (1990) found that certain rapid germination (high GE) barley varieties were susceptible to loss of Germination Capacity when stored in on-farm storage bins without means of aeration. This susceptibility seemed to be varietally dependent and could be predicted by storing harvested barley in sealed bottles at 131°F (55°C) for 3 or 4 weeks, then counting %GC. Healthy varieties showed >90%GC, whereas GC for varieties prone to dormancy dropped to 30–50%.

Gibson (1989) calculated that a minimum of 0.15 m3/min/metric ton (4.8 ft3/min/ton), well distributed throughout the grain, is necessary for adequate grain bin aeration.

LABORATORY BARLEY GERMINATIVE ENERGY (4 ML) RESULTS TABLE (PAULS MALT, 1998-2000)

24 HR. COUNT	48 HR. COUNT	72 HR. COUNT	INTERPRETATION FOR BARLEY
20-40	20-70	80-95	Probably still dormant, uneven germination
20-40	70-90	95+	Barley immature, slow germination
40-60	90-100	95+	Mature barley, rapid water uptake, good germination
70+	95+	95+	Very mature barley, germination difficult to control

A rapid tetrazolium reagent solution (1% w/v 2,3,5-triphenyltetrazolium chloride) method can also be used to measure GC, which, when applied to bisected grain halves, stains viable embryo and aleurone tissues bright red in 30 minutes under a vacuum.

GIBBERELLIC ACIDS: ALSO CALLED GA, GA1, GA3, GIBBERELLINS, GIB Gibberellins are a group of phytohormones first recognized in fungus-infected rice by Kurosawa in 1926. Later, it was noted that exogenous (originating outside of the cereal grain) preparations of gibberellins, GA3, from the fungus Gibberella fujikuroi could accelerate malting by enhancing the formation of the hydrolytic enzymes of cereals in the **aleurone layer**.

Exogenous GA is prepared as a white, crystalline powder and is dissolved and added to malting barley in steep water, after chitting, or by spraying during early germination in order to accelerate the malting process. Application via sprays in early germination is more efficient than steep water additions, requiring approximately 0.1 to 0.2 mg GA per kilogram of barley, compared to 1 to 2 mg/kg, when GA is added to steep soak water.

Uptake of exogenous GA can also be facilitated by the natural or mechanical process of **abrasion**.

The acceleration of modification, and subsequent time-savings, is the principal economic factor for the use of exogenous GA in the malting industry, although increases in fermentable sugars and fermentability have also been reported. However, due to the enhanced proteolytic activity of grains treated with exogenous GA, there is an increased supply of amino acids available to the embryo that accelerates the respiration and growth rate of the developing seedling shoot and root, which leads to higher malting loss and malt color formation in kilning.

It has been reported that all-malt beers from GA-treated malts may be more susceptible to development of Lactobacillus infection, presumable due to the higher amino acid content in the resultant beer.

Methods that limit proteolysis, such as additions of potassium bromate or the use of growth-retarding chemicals such as coumarin, organic solvents, ammonia, sulphur dioxide, acetic acid, copper sulphate, and formaldehyde, have been reported as co-applications with exogenous GA.

Endogenous (produced inside the cereal grain) GA1 and GA3 are found in all plants and are produced and secreted by the oxygenated embryo, inducing the synthesis and development in the aleurone layer of hydrolytic enzymes such as α-amylase, endopeptidase, limit dextrinase, and ribonuclease.

The development of other enzymes such as endo-ß-glucanase, endo-xylanase, cellobiase, and laminaribiase is stimulated by, but not dependent on, GA.

Enzymes of malting barley found primarily in the endosperm and, therefore, not affected by the presence of GA, include ß-amylase, carboxypeptidase, phytases, and lipases.

Recently researchers have shown that residual GA is found in malt in direct proportion to the amount added in steeping or germination; however no GA was found in finished beer.

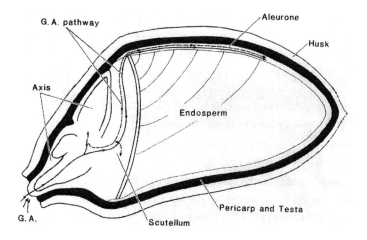

Fig. 37: Pathway of gibberellic acid (courtesy Palmer, 1989)

GLASGOW MALTMEN (THE INCORPORATION OF MALTMEN OF GLASGOW) The Glasgow Maltmen was formed in 1550 and is one of the oldest craft organizations in the malting industry. Maltmen not only made malt, they made beer and paid attention to "quality." The Maltmen now exists as a charity, making small financial awards to the International Centre of Brewing and Distilling (ICBD).

GLUTEN Gluten is the generic name for proteins that are used in bread-making flour that become sticky during dough conditioning, trapping CO_2 produced by yeast activity in baking, and giving volume to risen doughs. Glutens are the most abundant grain storage proteins, collectively known as prolamines or prolamins, from wheat (glutenins and gliadins), barley (hordeins), maize (zein), rye (secalins), sorghum (kafirin and sorghumin), and oats (avenins).

Prolamines are so-named because they have a very high proportion of the water-insoluble imino acid proline and amino acid glutamine, which can make them difficult to digest. Prolamines are ethanol-soluble so they are found in beer at trace levels, measured in parts per million.

Even at low levels, wheat glutenins and gliadins, rye secalins, and barley hordeins can cause intestinal disorders in a small percentage of people with specific allergies or sensitivity. Some confusion exists about the duplicate use of the term "gluten" in the bread dough/cereal milling context and when discussing human dietary ailments. In all dietary and disease discussions, the term "gluten" is only used to describe those reactive prolamines from wheat, rye, and barley, not the prolamines (or glutens) from maize, sorghum, oats, or any other cereal.

GLUTEN-FREE OR REDUCED GLUTEN Naturally gluten-free cereals like rice, corn, oats, sorghum, quinoa, teff, and millet can be used raw or malted to add unique flavors or as part of a "reduced-gluten" or "gluten-free" food or beverage. These products can be labelled "gluten free" only if there is zero chance of cross-contamination with even single kernels of wheat, rye, or barley that contain gluten. Cleaning, malting, or storing gluten-free grains in a facility or processing with equipment that also handles gluten-containing cereals eliminates the legal ability of the farmer, maltster, brewer, and food processor to label the grain and any subsequent product made from it as "gluten free."

A 2012 Mayo Clinic survey concluded that about 0.7% of Americans have celiac disease (CD), an autoimmune disorder that causes the body to attack the small intestine when gluten is ingested. Of those patients with CD, it is estimated that 50% of adults and 90% of child celiacs do not exhibit any symptoms or internal discomfort. Another 6% of the population is believed to have gluten sensitivity, a less severe problem digesting these proteins.

The percentage of households (20% in 2013) reporting purchases of gluten-free food products as general dietary improvement is about three times the number of households with diagnosed gluten sensitivities or allergies, and growing—making this a burgeoning market opportunity for reduced or gluten-free malted cereals.

In dollars and cents, sales of gluten-free products were expected to total $10.5 billion in 2013, according to Mintel, a market research company, which estimates the category will produce more than $15 billion in annual sales in 2016.

Guerdrum and Bamforth (2012) used an enzyme-linked immunosorbent assay (ELISA) method to track levels of prolamin through the brewing process. They found levels of prolamin in fermented beer (131 mg/kg) some 98% lower than the starting malted barley prolamin content (6,823 mg/kg). Prolamin in beer was undetectable after addition of a specific prolyl endoproteinase (PEP) enzyme preparation.

Recently, malt researchers (Tanner et al., 2014) showed that fungal (Aspergillus niger) preparations of PEP enzymes can virtually eliminate the immunoreactivity of celiac T-cells toward gluten peptides, thereby reducing the adverse effects of dietary gluten in some, but not all, situations. These authors however, were unable to test the efficacy of beer made from PEP-treated grains because, "In order to achieve this, subjects would have to drink 10 L of an average beer (at 100 ppm) per day to consume sufficient hordein (1 g) for a useful short term challenge."

On May 24, 2012, the Alcohol and Tobacco Tax and Trade Bureau (TTB) issued ruling 2012-2, "Interim Policy on Gluten Content Statements in the Labeling and Advertising of Wines, Distilled Spirits, and Malt Beverages regarding products containing gluten", finding that:

- The term "gluten-free" is considered misleading when used in the labeling and advertising of alcohol beverages to describe an alcohol beverage product that is made with any amount of wheat, barley, rye, or a crossbred hybrid of these grains, or any ingredient derived from these grains.
- Any industry member making a "gluten-free" claim on a label or in an advertisement to describe a product that is not made with wheat, barley, rye, or a crossbred hybrid of these grains, or any ingredient derived from these grains, is responsible for verifying that the producer has used good manufacturing practices to ensure that its raw materials, ingredients, production facilities, storage materials, and finished products are not cross-contaminated with gluten. Industry members are responsible for ensuring that

any gluten-free claim is truthful and accurate and should be prepared to substantiate such claims upon request. TTB may request samples to be submitted to TTB's Beverage Alcohol Laboratory to analyze the finished product.
- That labels and advertisements may include truthful and accurate statements that a product was "[Processed or Treated or Crafted] to remove gluten" for 6 - TTB Ruling 2012–2 OPR: RRD products that were produced from wheat, barley, rye, or a crossbred hybrid of these grains, or any ingredient derived from these grains, and then processed or treated or crafted to remove some or all of the gluten under the following conditions:
- One of the following qualifying statements must also appear legibly and conspicuously on the label or in the advertisement as part of the above statement:
  "Product fermented from grains containing gluten and [processed or treated or crafted] to remove gluten. The gluten content of this product cannot be verified, and this product may contain gluten."
  OR,
  "This product was distilled from grains containing gluten, which removed some or all of the gluten. The gluten content of this product cannot be verified, and this product may contain gluten."

(2) TTB will not approve labels containing the above claims unless the label application contains a detailed description of the method used to remove gluten from the product and R5 Mendez Competitive ELISA assay results for the finished product showing a response of less than 20 ppm gluten (and the name and manufacturer of the assay). Industry members should also be prepared to substantiate advertising claims with the same information, upon request.
- Statements characterizing the relationship of the product, or any substance within the product, to a

disease or health-related condition (such as celiac disease) are prohibited unless such statements comply with the requirements for specific health claims as set forth in the TTB regulations.
- TTB will approve labels with the statements described above only if TTB concludes, based on the totality of the information submitted, that the statement is truthful, accurate, and not likely to mislead consumers.

At least one new craft malthouse, Grouse Malting and Roasting Company (Wellington, CO) has dedicated its entire facility to malting only gluten-free grains (buckwheat, millet, quinoa, etc.), thereby ensuring that products are not cross-contaminated with gluten-containing grains. On their website Grouse malting says, "Grouse believes people have the right to great tasting truly gluten-free beers and is here to serve this industry."

Fig. 38: William Soles and Twila Henley of Grouse Malting and Roasting in front of their Germinate-Kiln vessel (GKV).

Fig. 39: Beautiful 122-year old wooden grain silo is the new home of Grouse Malting and Roasting in Wellington, CO.

GMO (GENETICALLY MODIFIED ORGANISMS) Whereas the U.S. biotech industry (maize, soybean, sugar beets, canola, Hawaiian papaya, alfalfa, and cotton) is currently valued at an estimated $200 billion dollars per year, there are no GMO or transgenic malting grain cultivars. It is anticipated that by 2016 there will be GMO rice varieties and by 2020, GMO wheats on the market. Stated rationale for GMO improvements to grains are improved pathogen, stress, disease, and herbicide resistance, as well as higher value food nutrition and production of biofuels.

GRADING (FROM ASBC METHODS OF ANALYSIS) The grading and certification of barley is a function of the U.S. Department of Agriculture (USDA); Grain Inspection, Packers, and Stockyard Administration (GIPSA); and, in Canada, of the Canadian Grain Commission.

Based on the properties of the grain, barley is divided into subclasses: Barley, Six-rowed Malting Barley and Six-rowed

Blue Malting Barley, and Two-rowed Malting Barley. Considering only the malting grades, six-rowed malting barley grades U.S. No. 1, U.S. No. 2, and U.S. No. 3 are defined on the basis of minimum limits for "test weight per bushel and for sound barley" and maximum limits for "damaged kernels, foreign material, skinned and broken kernels, thin barley, black barley, other grains, smutty, garlicky, ergoty, tough, injured by frost, injured by heat, dockage, presence of unsuitable malting type, and presence of two-rowed barley."

Barley that does not meet requirements of the three specified grades for subclass "Malting" is classified as "barley," that is, nonmalting grade.

Within any grade of malting barley, one can anticipate discounts for such factors as "high moisture (above 13.5%), high protein (above 13.5% dry basis), deficient % plump, excessive % thin, and gray, weathered, sprouted, or green" kernels.

In a similar fashion, the malting grades of two-rowed barley are classed as "U.S. No. 1, and U.S. No. 2," with U.S. No. 1 grade being high in bushel weight and in sound barley and low in other varietal types, thin barley, skinned and broken kernels, and in injury due to heat, frost, or molds.

Substantial emphasis is placed on varietal integrity. Acceptable malting varieties are specified for classification as white aleurone and blue aleurone six-rowed barley, and as two-rowed barley.

The Canadian Grain Commission uses the classifications No. 1 and No. 2 Canada Western (C.W.) Six-Row, and No. 1 and No. 2 Canada Western (C.W.) Two-Row. Barleys not meeting the malting grade specifications become designated as Nos. 1, 2, or 3 Feed.

In the U.K., in addition to grading, grain growers voluntarily commit their crop to and become certified in programs like the Assured Combinable Crops Scheme (ACCS) (www.assuredcrops.co.uk). Farmers must be able to show that they follow high standards from seed planting until the grain leaves the farm including:

Seed production—the grower must either provide details of certification for purchased seed or records of farm-saved seed including the qualifications for the person who treated the seed.

Pesticide and fertilizer application—Use of crop sprays are subject to detailed regulations including licensing of those who operate the spraying machinery, and regular inspections of the spraying machines.

Grain harvesting—quality inspections must be carried out on the combine harvester.

Grain storage—Detailed provision is made concerning the storage of the grain if it is stored on the farm, including buildings, rodent, insect, and moisture control.

Grain transportation—Vehicles used to convey grain have to undergo detailed inspections.

When malting grain leaves a farm, it takes with it a "stamped passport" that shows all the details regarding which farm it came from and that it is ACCS certified.

GRAIN BIN STORAGE AND WORKER SAFETY Pennsylvania State University extension service has found that most grain entrapment accidents develop around poorly flowing grain resulting from some type of grain spoilage issue in the bin. So if you can prevent spoilage, you may be able to eliminate the leading cause of bin entry and accidents. (http://www.grainbinsafetyweek.com/topics/proper-grainmanagement#sthash.Rdjxjj07.dpuf)

Steps that preserve grain quality and reduce grain bin accidents are:
- Clean grain: Prior to placing grain into storage, it's important to remove chaff, weed seeds, bees wings, and broken and immature kernels.
- Dry grain: To help ensure grain quality, dry grain to the recommended storage moisture before placing into bin. Kernel temperature should be <110°F (43°C) to preserve germination and <140°F (60°C) to preserve milling quality.
- Handle grain gently: Damaged, cracked, or broken kernels are more prone to deterioration and microbiological action than good, quality grain.
- Aerate grain: When storing grain after harvest, re-

duce temperatures to 50°F (10°C) or less as soon as possible by drawing cooler, night-time air through it. For optimal winter storage, uniformly cool grain to 30–35°F (–1 to 2°C) or less and maintain low grain temperatures for as long as possible to eliminate hotspots.

GRAIN WASHER   An overflow tank or flowing water stream device used prior to traditional steeping which removes dust, foreign kernels, stones, and loose microflora. When chemicals are added to grain washer water, then grain microflora can be reduced further via **alkaline steeping**.

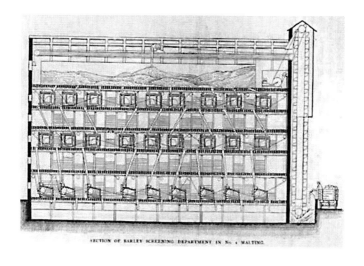

Fig. 40: Great Grimsby Maltings screening department (Barnard, 1889)

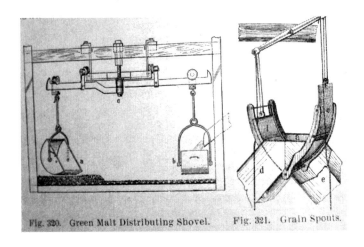

Fig. 41: Green malt distribution shovel and grain spouts.

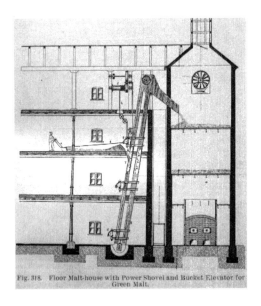

Fig. 42: Floor malthouse with power shovel and bucket elevator (Wahl and Henius, 1908).

GREEN MALT The name given to immature, wet malt after completion of the germination phase and before kilning begins. In malt whiskey distilleries with accompanying malthouses, or grain distilleries, green malts may be mashed directly to prepare "wash" or "green beer" with very high levels of enzymes and fermentable extract. Mashing with unkilned green malt is not done in breweries, even those with their own malthouses, because of the undesirable raw grain aroma and flavor notes persisting through to the finished beer.

GUMS An outmoded word used to describe the water soluble **pentosans** and **ß-glucans** in barley.

GUSHING The phenomenon of beer rapidly foaming or "jumping out" of a bottle or can of beer immediately upon opening. Malts made from grains that have been infected with molds, primarily of the Fusarium species, can cause extreme gushing, even to the extent of emptying the entire contents of the package. It is believed that molds produce hydrophobic proteins or hydrophobins, which have been shown to cause gushing at concentrations as low as 3 μg/ml.

There are several published methods for testing the ability of ground malt to produce gushing by adding a small aliquot to beer in a bottle, which is re-sealed, pasteurized, and shaken for three days prior to opening. Beer loss by gushing is measured by weighing contents before and after the test.

The same Fusarium species that can cause gushing can also produce **nitrosamines**.

H

HAACP (HAZARD ANALYSIS CRITICAL CONTROL POINTS) HAACP is a management system that identifies, documents, controls, and eliminates potential consumer hazards at their point of origin. Examples of potential hazards to consumer health in malting operations and products might include presence of **mycotoxins, nitrosamines**, pesticides, or heavy metals. HAACP programs provide a level of regulatory and insurance cost mitigation, and when coupled with an established quality management system, for example ISO9000 programs, that document process traceability, can help protect ingredient or consumer-product supplier's brand image and product liabilities.

HAZE, CHILL HAZE, AND PERMANENT HAZE Visible haze formed in beer when it is chilled to 32°F (0°C) and disappears when the beer is re-warmed to 68°F (20°C). Chill haze is formed between soluble proteins and polypeptides from malt and **polyphenols** (tannins) from malt and hops. Permanent haze is that which is still visible (not re-dissolved) when the beer is warmed and is formed by oxidation and polymerization of bound protein-polyphenol complexes.

HARTONG EXTRACT PROCEDURE A more involved malt extract method used by the Middle European Brewing Technology Analysis Commission (**MEBAK**) that seeks to gather more detailed information from malts extracted at four different isothermal temperatures: 68°F (20°C), 113°F (45°C), 140°F (60°C), and 176°F (80°C).

HEMICELLULOSE A general and archaic name given to groupings of arabinoxylans, pentosans, and pectins found in plant cell walls, accounting for about 10% of plant dry weight. Hydrolysis of hemicellulose by malt enzymes such as esterases, endoxylanases, and arabinoxylanases produces gums that are further degraded to oligosaccharides and simple sugars.

HIGH-DRIED MALTS Specialty malts that are produced in a standard kiln at high curing temperatures are often called "high-dried malts" in the U.S., or "color malts" in Europe.

HOMOGENEITY The homogeneity, or evenness, of malt modification is not normally assessed by standard milled malt sample extraction methods. Single kernels of barley weigh approximately 0.04 g, which means that a 50 g milled barley malt sample is made up of 1,250 individual kernels.

Therefore, the homogeneity or statistical distribution of individual kernel characteristics cannot be easily determined by standard malt analysis methods.

Other methods that measure the homogeneity of single kernel hardness, stealiness, mealiness, growth, and modification include **chitting** and **acrospire** growth counts, **Friability**, the Palmer half-kernel washing method (Palmer, 1999), or the Perten Single Kernel Characterisation System (SKCS).

HORDEIN Hordeins make up 20–40% of the proteins of barley and have structures similar to the **glutens** of wheat. It is not yet clear whether barley hordeins can cause intestinal upsets in the same way that wheat glutens do, but most physicians err on the side of caution and warn people with celiac disease or gluten sensitivity to likewise avoid or reduce consumption of all products made from barley, including beer.

Hordeins and their metabolites are found in beer and are involved in head retention and chill haze formation.

HUSK, ALSO CALLED HULL, GLUME, OR CHAFF The outermost layer of hulled cereals, composed of two leaf-like structures, the lemma (dorsal surface) and palea (ventral surface). The husk represents about 10% of total barley grain weight and protects the kernel and embryo from birds, insects, disease, and human handling during harvest, cleaning, malting, etc. The husk also provides much-needed matrix support during mash run-off and extract filtration in brewing.

Huskless or hulless or naked grains are more susceptible to damage in the field and in malting process handling.

HYDRATION Hydration is the distribution of water within the grain. It influences every aspect of cereal grain biochemistry, from grain development to storage after harvest, to viability and germination of the grain during malting. The functionality of endogenous grain enzymes on their substrates is called enzymic "hydrolysis," because it cannot occur in the absence of water.

Although the imbibition of water by cereal grains is

effectively complete at the end of steeping, some malting methods, particularly pneumatic compartments with high or dry air throughput, require additional water application to maintain adequate grain hydration throughout germination.

I

INTERNATIONAL BARLEY GENOME SEQUENCING CONSORTIUM (IBSC) A consortium (website barleygenome.org) formed by barley researchers from Australia, Finland, Germany, Japan, U.K. and U.S. that characterized the whole genome of barley, *Hordeum vulgare*, published in *Nature* on October 17, 2012. The barley genome is twice the size of human genome. Knowledge of the complete genome will help facilitate the development of new and better barley varieties able to cope with the demands of climate change. It should also help in the fight against cereal crop diseases, which cause millions of dollars in losses every year.

INTERNATIONAL CENTER FOR BREWING AND DISTILLING (ICBD) The Heriot-Watt University center for malting, brewing and distilling education and research located in Edinburgh, Scotland for teaching the malting, brewing, and distilling disciplines at diploma, undergraduate, and post-graduate levels. (website www.sls.hw.ac.uk/research/internation-centre-for-brewing-distilling.htm).

> The process of malting is described by Isodorus and Orosius, who wrote about the year A.D. 410, thus:
>
> "The Celtic nations steep their grain in water and make it germinate; they then dry and grind or pound it; after which they put it into water or infuse it. The liquor is fermented when it becomes warm, pleasant, strengthening, and intoxicating." (White, 1860)

Fig. 43: Loading the kiln with throwing machine at Laphroaig, Islay, Scotland.

K

KILNING After increasing the grain moisture in steeping and maintaining controlled moisture during germination, the moisture is then removed from the grain by kilning. Kilning is necessary to stop biological activity as soon as the required endosperm modification and enzyme levels are reached, to allow safe storage of the malt at low moisture content that also facilitates milling later, and to develop desirable flavor and color compounds in the malt while driving off some undesirable volatile compounds.

For maximum efficiency of kilning, the grain bed must be dead level and of uniform compaction. Kiln loading is usually stated in original grain steep-in weight units and should be between 70 and 100 lbs/ft2.

The early part of kilning is called the free-drying stage because unbound water is driven from the grain. This water causes evaporative cooling of the off-grain air so it is a little warmer than ambient but not as high as air-on temperature. Embryo respiration, endosperm modification, and enzyme synthesis continues during the free-drying stage and on until internal malt temperature reaches about 120°F (50°C).

Free-drying continues until grain moisture reaches 20–25%, at which time, the "break point" is reached when free water has been removed causing evaporative cooling to cease and air-off temperature to "break" or jump up to about

140–150°F (60–65°C). Thermal efficiency and water removal rates increase with higher air-on temperatures as do risks of heat damage to green malt, especially enzymes.

The second stage of drying is by diffusion as the water moves from the interior to the exterior surface, slowly raising the grain and air-off temperature, and ending at about 10–12% grain moisture. This stage is the most critical for thermolabile enzymes which begin to denature in high moisture, high temperature conditions. As the malt temperature increases, biosynthesis slows, resulting in buildup of low molecular weight sugars and amino acids that can be measured by the **Hartong cold water extract** method.

The third stage of kilning, called curing, must drive off water that is "bound" in the tissues of grain by higher air-on temperatures, ending with 3–6% final grain moisture and simultaneously produce color compounds—melanoidins—from the reaction of reducing sugars with amino acids.

Over-modification of malt as measured by higher FAN and soluble protein, and stewing grains while wet, yield higher melanoidin formation and higher malt extract color. In traditional floor maltings, grains were allowed to dry out for a period, called "withering," to complete modification and allow surfaces of grains to dry somewhat, preventing stewing, keeping malt color low, and reducing kiln power needs.

Important final malt flavors are developed during kilning from the enzymic and chemical oxidation of unsaturated fatty acids; Maillard reactions of amino acids; and sugars and thermal degradation of many flavor precursors, such as S-methyl methionine which can oxidize to form cooked corn flavors of **dimethyl sulfide** (DMS) (Bathgate, 1973).

The maltster's art in kilning is balancing grain modification and moisture levels with amount and timing of heat applied. Creation of darker malt colors or more intense roasted, caramel, or burnt flavors always means more denaturation of malt amylolytic enzymes. Even in pale malts, gentle kiln cycles will reduce Diastatic Power (DP) by as much as 50% from initial green malt activities.

KILNING TECHNOLOGY In 2012, a team led by Dr. Lindy Crewe of the University of Manchester, England, excavated a 2 m mud-plastered Cypriot Bronze Age malt kiln at the settlement of Kissonerga-Skalia near Paphos, Cyprus. In August 2012, an experimental archaeology team, led by Ian Hill of HARP Archaeology, recreated the drying kiln and recipe using traditional techniques for making a Kissonerga-Skalia Pale Ale:

- Using a porous sack, soak 1.5 kg of the barley grains in cool running water for 24 hours. A fresh running stream is ideal.
- Drain the grains and remove any unwanted stalks etc.
- Spread the grains evenly inside a semi-porous container (a shallow pottery vessel or wooden bowl for instance) and cover the container with a damp cloth and place the container out of direct sunlight.
- Uncover every 6 hours to stir the grains to avoid overheating and molding, and repeat until germination phase is complete (usually 3–4 days).
- When the grains have germinated, they are ready to malt. Split open a grain to check the germination, once the inner shoot of the grain has grown to around 75% of the length of the grain they are ready.
- To malt the grain, place in open containers and put them into the bottom of your drying kiln; once positioned, fire up the kiln and maintain a steady fire for 24 hours. This will produce a steady temperature of around 60°C for malting your grains to produce a pale malt.

More recently, malt kilns are built with single, double, or even triple perforated decks that allow the deep-bed germinated grain to be spread out in thinner layers for faster drying and curing. Some multi-floor kilns are operated statically—that is, the grain stays on a deck for the entire kiln cycle, or they can be operated sequentially by using tipping trays, augers, or screw conveyors to transfer grains from one deck to the other.

Kiln heat sources must prevent direct contact of the

combustion products with the malt at all times by using heat exchangers or non-combustion heaters. Kiln fan sizing is based on the average ambient relative humidity at the malthouse location, humid air requiring significantly more airflow than locales or seasons of dry ambient air.

Kilns should be built with intake and exhaust air mixing and heat exchange (i.e. glass tube) capabilities to recover and re-use as much exhaust heat as possible. Heating element bypasses are necessary to cool down malt at the completion of kilning and before transferring to storage.

Kiln loading for any configuration must be done with care to ensure that the grain bed is loosely packed and dead level, otherwise wet spots may require worker entry and manual pitchforking prior to higher air-on curing temperatures. Manually loaded kilns are usually more uneven than machine-loaded kilns and can require more airflow and energy. Kilns that are equipped with turning machines can usually deal with random wet spots by turning during drying, but may create their own "valley" and "hills" with different depths and packing densities, and may lengthen the drying process by mixing wet with already-dry grains.

Ringrose (1978) examined energy usage in large U.K. malthouses at different kiln loadings of 80–160 lb/ft2, and found that deeper kiln beds required six times as much energy (272 kwh/ton) than that with half the loading depth or twice the surface area. Mechanical methods of kiln loading—grain thrower vs. screw (auger)—made a difference too, with the former using 50% more energy than the latter at equal loading bed levels.

Pneumatic kilns can be built with either updraft or downdraft air flows. Both are acceptable, creating high-pressure zones on the air-on side of the grain bed and low pressure and dust/moisture cleaning issues on the air-off or exhaust side of the bed.

Updraft kilns have an advantage in allowing the identification of undried wet spots in the grain bed, either by visual observation or by comparing cooler thermocouple/thermometer readings. These wet spots are then easier to break up with a pitchfork or rake because they are on the top of the bed.

There still exist a few natural-draft kilns found in floor maltings on the islands of Orkney and Islay in Scotland. Generally, natural-draft kilns are built with a mechanical grain-spreading auger onto a wedge-wire floor with nominal 50% open space to a grain depth of 1–3 feet (0.3–1.0 m).

Drum roasters used in **specialty malt** kilning can vary in size from a few hundred pounds up to 3 tons of malt charge. To prevent sticking and burning of contents during operation, grain should be evenly distributed along the entire drum length at filling, the drum should rotate at 20 to 25 rpm, external heat applied evenly, and the design should ensure quick and complete emptying of the roasted malt as soon as desired color is reached.

Fig. 44: Nailsworth Maltings kiln chimneys (Barnard, 1889).

KOLBACH INDEX (S/T PROTEIN, SOLUBLE NITROGEN RATIO (SNR)) The Kolbach index is also called the Soluble/Total (S/T) protein or Soluble Nitrogen Ratio (SNR) measuring the water-soluble nitrogen or protein as a percentage of total nitrogen or protein. Target Kolbach indices vary for malts depending whether the mash will be 100% malt or malt/adjunct brews and whether decoction, programmed, or infusion mashes with or without protein rests will be followed. Kolbach indices for brewing malts range from 36–44%.

The Brewers Association (2014) believes that desired S/T protein level of 35–45% should be

"Achieved via varietal change, not by pushing germination moistures lower on existing varieties. Most group members favored 42% maximum; a few favored 45% maximum. Keep moisture levels elevated (43–44%) towards end of germination, with a caution against moderating S/T at the expense of higher Beta Glucan/Viscosity."

KJELDAHL MEASUREMENT OF PROTEIN A method for measuring total organic nitrogen in grains developed by Johan Kjeldahl at Carlsberg in Copenhagen in the 1880s. The method can be quite dangerous, because it involves suspended racks of boiling sulfuric acid. It has largely been replaced by safer Dumas and NIR procedures.

L

LIMIT DEXTRINASE (α-DEXTRIN 6-GLUCANOHYDROLASE) ALSO KNOWN AS R-ENZYME, PULLULANASE, ISOMALTASE, OR OLIGO-1,6-GLUCOSIDASE Limit dextrinase is a diastase produced in the aleurone layer during germination and cleaves the α-1,6 linkages in amylopectin, glycogen, and pullulan. Limit dextrinase works in tandem with α-amylase and ß-amylase to hydrolyze soluble starch molecules into fermentable sugars and dextrins in the mash.

LODGING The percent of grain straws that fall over in the field is called percent lodging. Lodging can interfere with mechanical harvesting, lowering crop yields. Increasing straw strength to minimize lodging is an important agronomic characteristic in plant breeding programs.

LOOSE SMUT *(Ustilago hordei)* A fungus that infects open cereal grain flowers in the field, producing a dark brown mass of spores that can replace then entire plant head. Fungicides are available for pre-planting seed treatments. Diligent rogueing of the field (walking through and pulling infected heads before they spread the disease) can also control loose smut.

MAFFEI MALTING SYSTEM A mechanized floor malting system developed in Germany in the 1920s that used moveable spouts to discharge grains out of steeping onto a semi-circular germination floor. Loading and emptying kilns was also mechanically aided.

MAILLARD REACTION The browning reaction between amino acids and reducing sugars in malt kilns that form **melanoidins** discovered by Louis-Camille Maillard. Many Maillard reaction products have sweet, caramel flavors, though some formed from diacetyl, acetol, and pyruvaldehyde can undergo Strecker degradation to form aldehydes and ketones with objectionable flavors.

> "This is grain, which any fool can eat, but for which the Lord intended a more divine means of consumption. Let us give praise to our maker and glory to his bounty by learning about... beer!"
> —*Friar Tuck, in the movie* Robin Hood, Prince of Thieves, *1991*

MALT The word "malt" probably derives from the Anglo-Saxon words metan, meaning to melt or dissolve, in reference to the softening of grain that occurs during malting, or malled, which means broken into fragments (Boulton, 2013) and is also the origin of the word "mallet" for a wooden hammer.

The word "maltster" is a contraction of "malt-stirrer," suggestive of the manual act of raking, shoveling or plowing the grain on a malt floor.

Any cereal grain that has been malted can be called malt. A wide variety of cereal grains and malt types are addressed individually in this text.

In some cases, for example loose interpretations of the **Reinheitsgebot**, or German beer purity law, a cereal grain that has passed through a malthouse with brief steeps, germination, or kiln cycles, have been called malts. Functionally, however, these "malts" must be thought of more as unmalted adjuncts than enzyme-rich malts, when they are used in breweries or distilleries.

COMPARISON OF TWO-ROWED AND SIX-ROWED BARLEY AND MALT (FROM SCHWARZ AND HORSLEY, 2012)

TYPICAL BARLEY & MALT ANALYSIS	TWO-ROWED BARLEY	SIX-ROWED BARLEY
Grain yield (bu/acre)	145 (Western, irrigated), 90 (Western, dryland)	88 (Midwestern, dryland)
Test weight (lb/bu)	54 (Western, irrigated) 52 (Western, dryland)	47 (Midwestern, dryland
Kernel plumpess (% >2.4 mm)	89 (Western, irrigated) 78 (Western, dryland)	78 (Midwestern, dryland)
Total protein (% dry basis)	11.5	12.5
Malt extract (% dry basis)	81	79
Soluble protein (% dry basis)	5.0	5.5
Diastatic Power (°Lintner)	120	160
Alpha-amylase (D.U.)	50	45
Wort viscosity (cP)	1.5	1.5
Wort ß-glucan (ppm)	110	140
Wort color (SRM)	1.5	1.5

Brewing malts

Lager/Pilsner malts—Two-rowed spring barley malts with low protein (9–11%), well-modified, with low malt extract color of 1–3 SRM (2–6 EBC), produced by 42–44% moisture at steep-out, 59–65°F (15–18°C) germination temperatures and initial kilning temperatures of 104–122°F (40–50°C), followed by an intermediate drying phase of 149–167°F (65–75°C), and final roasting temperatures of 175–185°F (80–85°C).

Lager/Pilsner malts may be used at 100% of the grist to produce low color lagers, pilsners, and light beers, but also has sufficient amylolytic enzyme content to be brewed with up to 40% un-malted adjuncts or dark malts.

Pale ale malts—Two-rowed spring or winter barley malts, well modified and kilned similarly to lager/pilsner malts.

> "It is supposed that the art of malting was introduced into this country by the soldiers under Julius Caesar, 55 b.c. Vinegar and beer were the ordinary beverages of the Roman soldiery. The former was made very strong, and drunk, diluted with water, when on their march. Amongst the numerous features of civilization brought into this country by that energetic and martial people, none appear to have been so thoroughly appreciated as the introduction of a new and easily prepared intoxicating drink." (White, 1860)

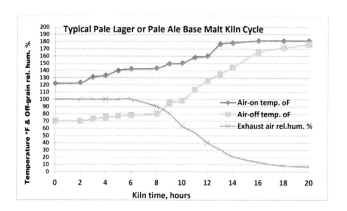

Fig. 45: Typical pale lager or pale ale base malt kiln cycle.

Distillers' malts The production of distilled spirits in the U.S. requires certain quantities of malted and unmalted cereals as defined by the U.S. Code of Federal Regulations:
- Malt whisky mash must contain at least 51% malted barley.
- Wheat whisky mash must contain at least 51%

wheat.
- Rye whisky mash must contain at least 51% rye.
- Rye malt whisky mash must contain at least 51% malted rye.
- Bourbon whisky mash must contain at least 51% corn (maize).
- Corn whisky mash must contain at least 80% corn.

Priorities for making malts for distilling are maximization of fermentable extract through endosperm modification and development of amylolytic enzymes in germination, and protection of the same in the kiln. Because there is no wort boiling in distilling as there is in brewing, many malt enzymes remain active throughout mashing and into fermentation, resulting in an estimated 20% increase in total alcohol yield due to continued conversion of starches to fermentable sugars.

Breeders of barley for distilling (mostly in Scotland) focus on the following, as described recently by Brown et al. (2011): greater agronomic and extract yields, rigid straw, shatter-resistant ears, enhanced disease resistance, improved grain size and uniformity, and ease of malting.

Gentle **kilning** conditions that allow continued and warm germination, like those for distilling malts, can actually produce dried malts with higher α-amylase and peptidase levels than found in the green malt, slightly lower ß-amylase, limit-dextrinase and lipase levels, and large reductions (80–100%) of thermolabile endo-ß-glucanase, maltase, peroxidase, catalase, and lipoxidase activity levels (Dolan, 2003). If the kilning is done too gently however, the possibility of higher grain moisture or "green grass" sensory notes in the final malt and product exist.

Kilning cycles for distiller's malts typically run 12 hours at 140°F (60°C), 12 hours at 155°F (68°C), and 6 hours at 162°F (72°C), with high airflows of 1,200–2,200 ft3/min/ton for rapid drying down to 5–6% grain moisture.

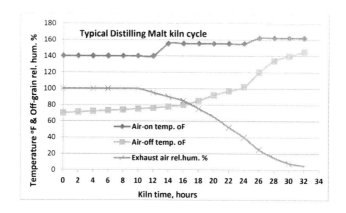

Fig. 46: Typical distillers malt kiln cycle.

> "And malt does more than Milton can,
> To justify God's ways to man."
> —*A. E. Housman*, A Shropshire Lad

COMPARITIVE ANALYSES OF COMMON MALTS (ADAPTED, THOMAS AND PALMER, 1994)

	MALT COLOR, SRM	MALT EXTRACT, FINE GRIND, %D.B.*	MALT ENZYMES, RELATIVE
BREWER'S BARLEY MALTS			
Lager (Pils) two-row malt	1–3	79–83	High
Lager (Pils) six-row malt	1–3	77–81	High
Pale/Mild Ale malt	2–4	81–83	Medium
Vienna malt	2–5	79–82	Medium
DISTILLER'S BARLEY MALTS			
Peated or unpeated Distiller's malt	1–3	80–83	High
High DP Distiller's malt	1–3	80–83	High
ROASTED BARLEY MALTS			

Amber malt	20–50	75–80	Low
Brown malt	50–100	70–76	None
Black malt	400–600	68–72	None
Caramel malt	6–120	78–80	Low
Caramel Pils malt	6–20	76–80	Medium
Chocolate malt	250–500	70–74	None
Crystal malt	40–200	50–65	None
Munich malt	5–30	78–82	Medium
ALTERNATIVE MALTS			
Wheat malt	1–4	84–88	High
Roasted barley	300–800	65–70	None
Rye malt	3–5	65–70	Low
Oat malt	3–30	60–72	Low

*Extracts (d.b. = dry basis) on color malts are usually run as adjuncts in the malt lab because that is how they are used in brewing; that is, color malts are mixed with brewer's malt and extract determined by difference from the brewer's malt extract.

High Diastatic Power (DP) malts—Two-rowed or six-rowed barley malts with high protein levels (>12%) and kilned at low temperatures to stabilize hydrolytic enzymes for mashing with adjuncts.

Peated distilling malts—Two-rowed or six-rowed barley malts, kilned at low temperatures with peat smoke, called "reek" in Scotland, while the malt is hand-dry (about 15–25% moisture) in the kiln to produce lightly (1–5 ppm total phenols), medium (5–15 ppm), and heavily peated (15–50 ppm) flavors, principally formed from malt phenols (Walker, 1990). Important flavor-active compounds from peat smoke include phenol, isomeric cresols, xylenols, and guaiacol. Residual amounts of these compounds depend on peat characteristics (seaside seaweed or inland grass bogs) and peat combustion temperatures in the kiln.

Non-peated distilling malts—Two-rowed or six-rowed barley malts, kilned at low temperatures without peat smoke.

MALT AGING Nearly every brewing textbook says that a minimum time for malt aging is required prior to milling for optimum mash extraction. Most of these books arrive at the actual aging time required through empirical, word-of-mouth brewing tales when improperly aged malts produced unexpected extract, mash filtration, fermentation, flavor, or beer haze problems.

A large U.S. lager brewer experiencing fermenter "hangups" discovered, during root-cause analysis, that 70% of the malt blend used to brew the faulty beer had been aged 14 days or less before milling in the brewhouse. This finding resulted in establishment of a new malt age minimum specification of 28 days.

Stopes (1885) claimed that

"The most satisfactory equilibrium of the constituents of malt is reached in a month or six weeks after drying, and it changes within two months.... Malt direct from the kiln, however carefully dried, is not suitable for immediate use in the brewery; as unless it be kept for from five to seven weeks in store, the worts produced from it do not break well on boiling, the fermentations are irregular, the yeast deteriorates, and the resulting beer remains cloudy for a considerable time."

Indicating that malt aging was (and still is) a controversial topic, Stopes quotes Trausing as saying "the age of malt has no influence whatever upon the quality of the beer," which Stopes calls a "very extraordinary statement."

Even Briggs (1998), considered by some as the most comprehensive contemporary textbook on malting, does nothing to advance the science of malt aging, only saying

"…while pale malts are believed to improve if stored for 1–3 months before being used for brewing… Kilned pale malt is usually stored for 4–6 weeks before it is used in brewing. There is some evidence, and a strong belief, that the quality and yield of extract obtainable from the malt improves during this storage period."

Though not widely done, the best service a malthouse can do for its customers, with regard to storing, aging, and rotating stock on-site, is to stamp the kiln date on every bag of malt.

Insect infestation and wet spots in grain and malt bins can cause ruined malt or, in severe cases, fire by spontaneous combustion. Temperature probes with local and computerized readouts and alarm capabilities (i.e. cell phones, iPads, etc.) are highly recommended.

MALT ANALYSIS—"THE ANALYSIS IS ONLY AS GOOD AS THE SAMPLE."

In 1759, The London and Country Brewer. A New Treatise on Liquor (Stopes, 1885) listed several techniques "To Know a Good from Bad Malt":

> "First, By the Bite; is to break the Malt-corn across between the Teeth, in the Middle of it or at both ends, and if it tasteth mellow and sweet, has a round Body, breaks soft, is full of Flour all its Length, smells well, and has a thin Skin, then it is good.
>
> Secondly, By Water; is to take a Glass near full, and put in some Malt, and if it swims, it is right, but if any sinks to the Bottom, then it is not true Malt, but steely and retains somewhat of its Barley Nature; yet I must own this is not an infallible Rule, because if a Corn of Malt is cracked, split, or broke, it will then take the Water and sink, but there may an Allowance be given for such Incidents, and still room enough to make a Judgement.
>
> Thirdly, Malt that is truly made, will not be hard and steely, but of so mellow a Nature, that if forced agains a dry Board, will mark, and cast a white Colour almost like Chalk.
>
> Fourthly, Malt that is not rightly made will be Part of it a hard Barley Nature, and weigh heavier than that which is true Malt."

Malt analysis of today is statistically proven by collaborative testing through organizations like the **American Society of Brewing Chemists (ASBC)**. The ASBC also organizes check services for malt laboratories, for a fee, which is advisable for periodic assurance of method accuracy. Many of these methods can fairly simply be done by craft maltsters themselves. (These are shown below with an asterisk.)

Alpha-amylase—Malted barley **base malts** should contain 45–80 DU (ASBC Dextrinizing Units, to have sufficient working power for mash starch hydrolysis).

Beta-glucan—Wort ß-glucan is the amount of hydrolyzed cell wall material that is warm water soluble. High levels (>150 ppm for malted barley) indicates under-modification or presence of unmodified grains (e.g. **dead barley** as measured by Friabilimeter) in the sample.

Color*—A sign of good protein levels, malt modification, and kilning practice, as well as the maltster's ability to blend lighter and darker malts together.

Conversion time, minutes*—The time that it takes for malt enzymes to gelatinize (swell), liquefy (reduce viscosity), and saccharify (make sugars from starches) its own starches in a laboratory mash as measured by the starch/iodine color reaction. Long conversion times (>7 minutes in ASBC mash) indicates either insufficient enzyme levels or under-modification of cell walls or protein matrix, or both. When extract conversion times were checked on isolated Friability fractions, Thomas (1986) found that Friable Flour, Half-glassy, and Case-Hardened malt fractions converted in normal 5–7 minute conversion times, whereas isolated Friability Hard-End fractions took >20 minutes and Dead Barley fractions >60 minutes to convert.

Diastatic Power—The combined activity of all amylolytic (starch-hydrolyzing) enzymes on a laboratory prepared starch solution. Diastatic Power (DP) is determined by grain type, variety, protein level, and kilning conditions. High DP malts convert starches rapidly, low DP (<120°ASBC) malts may not convert starches completely in the given time, resulting in reduced fermentability and alcohol levels.

Extract*—All of the starches, sugars, proteins, and lipids from milled malt that dissolve in warm water, which is predictive of extract and alcohol yield in brewing and distilling. Comparatively low extract indicates under-modification, higher than normal extract could indicate excessive malt damage and insoluble material (husk) loss during cleanoff, artificially elevating the percentage of other, soluble extract materials. Over-modification will reduce extract levels due to excessive botanical growth, heat, CO_2, and water vapor loss.

FAN—A measure of the proteolysis that occurred during malting and is available for yeast nutrition, after additional FAN production from proteolysis in mashing at lower, "protein-rest" temperatures. Desirable wort FAN levels for malt + adjunct brewing are 170–210 ppm, and for all-malt beers, 140–180 ppm.

Filtration time, minutes—A combination of extract, ß-glucan, and viscosity which may be predictive of mash filtration performance.

Fine-coarse difference*—The percentage extract difference between fine and coarse grinds in the malt lab, indicative of overall malt modification. F/C difference <1.5% is considered well modified.

Friability*—A rapid measure of both extent and homogeneity of malt modification.

Moisture*—The total amount of water in a milled sample of grain after kilning, usually 3–6%. If moisture is too low, excessive energy was used in kilning and malt husk may shatter during milling. High moisture can facilitate lipid oxidation, rancidity, insects, and mold growth, also reduces percentage malt extract, as is.

Whereas percentage moisture tells "how much," it does not tell "where" the moisture is in the grain. Immediately after kilning, malt with 5% average moisture, might, in fact, have 1–2% moisture husk and 7–8% moisture interior endosperm levels. Aging the malt for several weeks prior to milling allows this moisture gradient to equilibrate, facilitating extraction, enzyme attackability, and even fermentability of resultant worts.

pH*—For any given malt compared to the standard, lower pH may indicate over-modification or microbial growth, higher pH may indicate under-modification.

Protein—Higher grain protein means slower modification and lower total extract. Low grain protein might not produce enough enzyme activity for mashing. High soluble (>5.8%) and soluble/total (S/T) protein ratio (>48%) indicates over-modification which may negatively affect beer foam and haze stability and malt color, as well as making beer less microbiologically stable. Low soluble (<4.6%) and S/T protein ratio (<42%) may indicate under-modification, producing insufficient amino acids and peptides (short proteins) for healthy yeast nutrition.

Screening*—The fractional weight, using calibrated ASBC sieve screens, that measures the plumpness and size uniformity of raw grain and finished malt samples.

Viscosity*—High viscosity (>1.55 cps) of malt extract can be indicative of high levels of wort ß-glucan arising from general under-modification or the presence of a small percentage of dead barley or all-glassy fractions in the malt, as identified by Friabilimeter analysis. High malt extract viscosity may be predictive of poor mash filtration (run-off) performance.

Fig. 47: Basic malt lab instruments (l–r): sample bag sealer, stereo microscope, moisture tester, coffee cup, grain assortment screens, and digital scale.

Fig. 48: Well-equipped lab at Blacklands Malt (Austin, TX) with (l–r): 1) Analytical balance and moisture analyzer with kiln samples on table. 2) Glassware and lab tools cabinet. 3) Microwave oven for food or microwave color method. 4) Magnetic stir plate for malt extraction. 5) On floor: food and water for the malthouse dog! 6) Filter and Imhoff cone rack. 7) Coffee machine! 8) Double sink. 9) Manual coarse-grind mill clamped to table. 10) Motomco 919 moisture meter. 11) Burr fine-grind coffee mill. 12) Seedburo barley pearler for checking samples for pre-harvest sprouting.

MALT DAMAGE It goes without saying that the more malt is moved or overgrown in germination, the higher risk of damage. Highly modified malt and roasted, low-moisture malt is more susceptible to breakage and husk loosening. Dust generated during handling can create housekeeping, explosion, microorganism growth dangers, and higher malting loss.

Malt can be damaged by impact against loading equipment, bends in chutes or conveyors, elevator buckets, or kernels hitting other kernels or the bottom of deep bins. Damage by abrasion may occur on sides of chutes or shafts, by conveyors, elevators, or by kernels rubbing against kernels.

Malt damage may be reduced by decreasing the time and distance of free-fall from loading spout or conveyor to storage bin, increasing the cross-sectional area of the free-fall stream, using flexible extension hoses or tubes to maintain contact with the top surface of the transferred malt, slowing transfer equipment down, or minimizing the abrasion of grains against each other.

MALT EXTRACT In malt analysis, extract is simply a percentage of warm-water-soluble carbohydrate and protein and yeast nutritional materials per unit weight of malt. Malt extracts comprise of 90–92% carbohydrates (~75% of which are fermentable), 3–6% nitrogen compounds, 1.5–2.0% polyphenols, 1.5–2.0% inorganic materials, 0.2–1.0% lipids, and 0.04–0.07% vitamins.

Percentage extract is simply calculated as the total amount of dissolved material weight in the wort or extract divided by the total weight of raw materials.

"Dry basis" extract means water has been removed by calculation and is the extract figure used by maltsters to determine extent of malt modification and when comparing malts with different moisture contents.

"As is" extract can be used by brewers and distillers when buying malt ($/lb extract, as is) and when calculating expected brewhouse yield from a given charge of malt.

To convert from Institute of Brewing (IOB) units of liter degree per kilogram to American Society of Brewing Chemists (ASBC) percent extract units:

ASBC % Extract = 0.261 (IOB L°/kg)

Products are available called Malt Extracts, or Malt Syrups, which are sweet worts (no hops added) that have been concentrated in a multiple effect evaporator into thick syrup with 75–85% solids or dry powder by removing most or all of the water, and packaged for use by bakers and homebrewers. These syrups can be produced as non-diastatic (having now enzyme activity), or low to medium diastatic, generally in a range of colors, 130–500 SRM (260–1,000 EBC); and are used in baking white breads, rolls, and hamburger buns at 0.25–2.5%; in rye breads, English muffins, and crackers at 1–6%; and up to 18–20% in cookies (Pyler and Thomas, 2000).

MALTING The entire process of soaking, sprouting, drying, roasting, aging, and blending cereal grains to develop extractable material, enzymes, colors, and flavors suitable for use by brewers, distillers, and food companies, is called malting.

Historically, brewers and distillers made malt themselves, in tiny malthouses, purchasing locally grown grains. Today, Malteries Soufflet, the world's largest and family-owned malting company headquartered in France, produces five billion pounds of malt each year, more than 500,000 lbs of malt (250 tons) every hour of every day, which is equal to about 20 days of the total combined production of all current North American craft maltsters (Appendix C).

MALTING EDUCATION Many new-generation maltsters as well as seasoned employees of large malting or brewing and distilling companies have attended or plan to attend one of several schools that teach malting, such as:

- The three-day Malting Program or two-week Intensive Malting Program at the Canadian Malting Barley Technical Centre (Winnipeg, Canada), website cmbtc.com
- The two-week Brewing and Malting Science Course taught annually by the Master Brewers Association of the Americas (Madison, WI), website mbaa.com
- The four-day Barley Malt Quality Evaluation Short Course (co-sponsored by ASBC) and three-day Barley Production Field School at North Dakota State University Northern Crops Institute (Fargo, ND), website northern-crops.com
- Briess Malting Company's (Chilton, WI) annual two-day Malt and Brew Workshop, website briess.com
- The Fermentation Science course and UC Davis Extension Professional Brewing Program at University of California at Davis, website ucdavis.edu
- Diploma, M.Sc., and Ph.D. programs at the International Centre for Brewing and Distilling at Heriot-Watt University (Edinburgh, Scotland), website

icbd.hw.ac.uk
- The Arbeitstagung (Malting Technology Workshop) at Doemens (Gräfelfing, Germany), website doemens.org
- Technische Universität München at Weihenstephan, website wzw.tum.de
- Siebel Institute of Technology and World Brewing Academy (a Siebel Institute and Doemens Academy partnership), Chicago, IL and Munich, Germany; website siebelinstitute.com
- The Scandanavian School of Brewing, Copenhagen, Denmark, website brewingschool.dk

Most craft maltsters also depend on the Craft Maltsters Guild online discussion group (north-american-maltsters-guild@googlegroups.com) and their extensive and sometimes antiquated collection of barley and malt textbooks.

Fig. 49 (above): Great Canal Maltings Glasgow, Scotland. Fig. 50 (below): Malting floor at Great Canal Maltings Glasgow, Scotland (Barnard, 1889)

MALTING LOSS Conversion of any raw grain to dried, cleaned malt results in total malting losses of 10–20%. As a percentage of barley dry matter steeping losses are 0.5–1.5%, growth and clean-off of rootlets (culms) 2–5% and respiration losses of CO_2 and heat 3–8%, depending on type of grain, amount of kernel and husk damage, and germination conditions.

Recent research involves using microbial preparations (Lactobacillus plantarum AB1) and chemically acidified solutions (CAS) to reduce malting loss by as much as 75% while producing adequate malt modification and beer quality (Mauch et al., 2011).

MALTSTERS' ASSOCIATION OF GREAT BRITAIN (MAGB, WEBSITE UKMALT.COM) The principal U.K. malting trade organization founded in 1827 initially to help maltsters deal with the complicated (and bizarre?) U.K. tax laws that were in place for almost 200 years until late in the 19th century.

These antiquated tax laws were based on volume of malt measured in the couch between steeping and germination which determined the volume of all-malt beer produced and taxed owed. The malt tax was very important to the country's financing, accounting for nearly 10% of all tax collected by the U.K. government in 1868.

Today, the MAGB functions in a similar manner to the **American Malting Barley Association (AMBA)** in helping member companies ensure quality grain and malt supplies, interpretation and input into legislative processes, and maltster training and qualification.

MALTSTER'S THUMB "The maltster's thumb" describes the ubiquitous white stripe seen along the inside edge of the thumb which comes from picking up a single kernel of malting (germinated) grain and simultaneously pinching to burst and pulling down the endosperm contents from top of the thumb down to about the first knuckle. Cell wall ß-glucans and endosperm proteins in undermodified grains will roll up, like gum, producing a ball of material and no or minimal visible marks on the thumb. Well-modified grains will produce a white stripe on the thumb caused by endosperm starches which have been freed of their gummy cell walls and sticky protein matrix glue during malt modification.

Fig. 51: The Maltsters Thumb. (Photo courtesy Roger Putman.)

MARIS OTTER BARLEY AND MALT Maris Otter winter barley was first developed in 1958, quickly becoming the premier malting barley in the U.K. due to its ease of modification and flavor characteristics. It lost its popularity in the 80s and 90s because it yielded 30–40% less per acre than newer barley varieties.

The unique flavor of Maris Otter was demonstrated in a test at the Brewing Research International (Nutfield, England), reviving worldwide interest in the variety (Ponsonby, 2007) including companies like Sierra Nevada (Chico, CA) and Molson Coors (Burton-on-Trent, England). This coupled with higher contract barley and malt prices and a new marketing technique called the "vendor assurance" program helped save Maris Otter barley from certain marginalization to specialty barley status.

Today, it is interesting to note that, when the few craft brewers who deign to tell their customers about barley varieties, it is the mystique of Maris Otter that they boast about most often. Perhaps some U.S.-grown varieties, like the experimental "Full Pint" from the Pacific Northwest, will enjoy similar brewing flavor name recognition in the not-too-distant future.

According to Tom Blake, Senior Barley Breeder at Montana State University,

> "The reason we don't grow Maris Otter (or many other European-adapted barley varieties) in the U.S. is because they are late to flower, demand lots of water and require a very long growing season."

MASH CONVERSION IODINE TEST A rapid and useful brewhouse method for determining whether there are sufficient amylolytic enzymes present, and that all cereal starches (malt and adjuncts) are gelatinized (swollen) making them accessible to hydrolysis in the mash. Starch molecules greater than approximately 18 glucose units in length will show red, blue, or black coloration when a few drops of a solution of 2% potassium iodide are added to a spoonful of mash on a white, clean ceramic surface.

The method is often used as a go/no go determination prior to advancing to "mashing-off" or lautering operations in the brewhouse or distillery. **Amylopectin** shows a red-purple color in the mash conversion iodine test, whereas **amylose** shows a much darker blue-black color.

MASTER BREWERS ASSOCIATION OF THE AMERICAS, MBAA The Master Brewers Association of the Americas is a 501(c)(3) non-profit organization formed in 1887 with the purpose of promoting, advancing, and improving the professional interest of brew and malt house production and technical personnel, website mbaa.com.

MEBAK (MIDDLE EUROPEAN BREWING TECHNOLOGY ANALYSIS COMMISSION) The German association that standardizes and publishes analytical methods in malting and brewing.

MELANOIDINS Colored pigments (furfurals, isomaltol, maltol, and pyrazines) formed by Maillard browning reactions of sugars and amino acids in malt kilns. Analytical levels of FAN, amino acids and fermentable sugars in malt are reduced by melanoidin formation. In addition to brown, red, and black colors, melanoidins produce flavors of malty, caramel, toffee, coffee, and burnt toast in beers and whiskey.

MICRONIZED GRAINS Cereal grains (mainly barley, wheat, and maize) that are heated under infrared lamps while travelling on a conveyor belt in a micronizer unit. The grains dry, swell, and burst, like popcorn and the interior starches gelatinize, making them useable as pre-gelatinized adjuncts in brewing.

MICROSCOPIC EVALUATION OF MALTS Using light microscopy, Brown and Morris (1890) demonstrated that enzymic degradation of cell walls (cytolosis) is the first visible step in barley endosperm modification. This has been confirmed and expanded on more recently by high-resolution scanning electron microscopy (Palmer, 1999).

Light microscopy has also been used in conjunction with fluorescent dyes, like **calcofluor**, to monitor the presence and location of specific enzymes and polysaccharides within the barley endosperm, thereby monitoring enzyme movement and cell wall breakdown during malting.

An inexpensive binocular stereo dissecting microscope is useful in a malt lab in examining whole and cut sections of native and malted grain for mealiness (flouriness) and stealiness or glassiness (protein hardness) in the endosperm.

MILLET Members of the Poaceae family of grasses with very small (3–10 mm) seeds, including proso, pearl, finger, and foxtail millets in white, gray, yellow, or red varieties. In the U.S. it is grown mainly for bird seed although it is found in human foods in other Asian and African countries. There is recent interest in malting millet to make **gluten-free** beer.

Due to its size, millet grains absorb water very rapidly and care must be taken not to over-steep them. Warm germination temperatures of 70–75°F (21–24°C) for 2–4 days are required to malt millet.

Malthouse screens must be made very fine when malting millet and other small grains. One method of handling millet in malting tanks is to drill staggered 1/16 inch holes, approximately 1/8 inch apart in stainless steel sheeting resulting in 25–30% open area which is enough to retain these small grains, as well as restrict airflow a little for evenly distributed drying in kiln.

Traditional millet beer, called *kunuzaki, mbege, toguwa,* or *uji* is made in Africa by soaking millet and drying it in the sun, followed by grinding in stone mortars, fermenting spontaneously for 24 hours, then consumed while actively fermenting.

Fig. 52: Malted millet.

Fig. 53: Sample bags of malted millet from gluten-free Grouse Malting and Roasting Co., Wellington, CO.

MILLING Before malted or raw grains can be used in brewing or distilling, the particle size must be reduced by milling, or grinding, the desired sizes or milled particles depends on the type of grain/extract separation equipment used in the facility. For example, mash filters can handle the finest grind, even those produced by hammer mills, whereas lauter tuns need a percentage of large husk pieces released by two-, four-, or six-row roller mills to serve as the filtration matrix or bed.

Milling for traditional mash tuns at whisky distilleries is usually handled with two- or four-roller mills, yielding particle size distributions of 20% husk (>1.0 mm), 40% coarse grits (>0.59 and <1.0 mm), 30% fine grits (>0.25 & <0.59mm), and 10% flour (<0.25 mm). At breweries and distilleries with combination lauter/mash tuns or separate lauter tuns, slightly finer grinds through six-roller mills are typically employed. Steam conditioning of the dry husk prior to milling can help maintain filterable husk fractions while adjusting mill roller settings tighter for improved wort separation and extract yield. Mash filters generally are able to handle much finer hammer-milled malt flours.

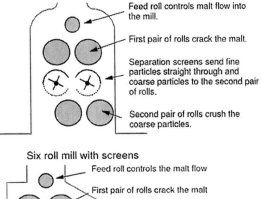

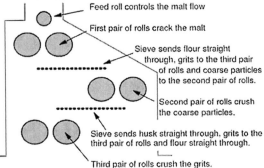

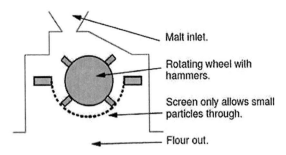

Fig. 54 (pg 123): Four-roll malt mill with screens. Fig. 55 (above): Six-roll malt mill with screens. Fig. 56 (below): Hammer mill with screens. (Diagrams courtesy Institute of Brewing and Distilling, London.)

MILLED MALTED BARLEY FLOUR PARTICLE SIZES SEPARATED USING ASBC SIEVES AND RECOMMENDED PERCENTAGE RETAINED FOR LAUTER A TUN OPERATIONS FROM THE AMERICAN SOCIETY OF BREWING CHEMISTS (ASBC) AND EUROPEAN BREWERY CONVENTION (EBC).

ASBC SCREEN NUMBER	NOMINAL PARTICLE SIZE, MM	MALT FRACTION	ASBC RECOMMEN- DATIONS, % RETAINED	EBC RECOMMEN- DATIONS, % RETAINED
8	>2.2	Husk		
10*	2.0–2.2	Husk	13	
14*	1.4–2.0	Husk & Grits	20	
15	1.3–1.4	Husk & Grits		18
16	1.2–1.3	Husk & Grits	30.3	
18*	1.0–1.2	Coarse Grits	32	7
30*	0.6–1.0	Coarse Grits	24	35
50	0.4–0.6	Fine Grits		
60*	0.3–0.4	Fine Grits	6	
70	0.25–0.3	Fine Grits		20
100*	0.15–0.25	Flour	2	8
Pan*	<0.15	Fine flour	3	12

*= ASBC recommended sieves

The most widely used mills are roller mills with gap-adjustable sets of two, four, or six rollers, with smooth or fluted surfaces, that work in tandem to crush, shear, and cut malts into the desired distribution of husk, grits, and flour.

Mechanical shakers are available (e.g. Ro-Tap) with ASBC sieves for around $2,000, or the same sieves can be shaken manually (ASBC method) to give accurate milled malt particle size data for around $300 total.

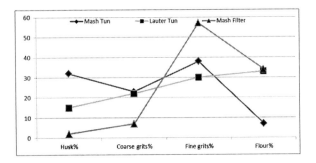

Fig. 57: Approximate malt sieve distribution for different mash filtration apparatus (mash tun, lauter tun, and mash filter).

MODIFICATION, OR MALT MODIFICATION The sum of the changes that take place in cereal grains during germination. Chief among these are the softening of the endosperm and the development of enzymes.

The biochemical and practical relationships between grain hydration, endosperm structure, and enzyme activity during cereal germination were first examined scientifically by H.T. Brown and co-workers in Great Britain (1890) and G. Haberlandt in Germany (1890). These independent malt researchers expressed different views concerning the characteristics and progression of modification in germinating barley. While delivering his *"Reminiscences of Fifty Years of Experience of the Application of Scientific Method to Brewing Practice,"* Brown (1916) predicted that "…chemists will not be united upon it in its bicentenary in 2014 a.d." (i.e. 200 years after Kirchoff discovered that "vegetable albumin" of grain converts starch to sugar).

Brewers and distillers commonly look for the most important measures of modified malt, including varietal purity, moisture level, percent extract (as is, dry basis, and fine–coarse difference), extract color, S/T protein ratio (Kolbach index), α-amylase and DP activity, total ß-glucan, FAN, friability, and fermentability.

The methods of determining malt modification can be grouped into three main types and are discussed further in this handbook; **microscopic evaluation, chemical analysis**, and **physical testing**.

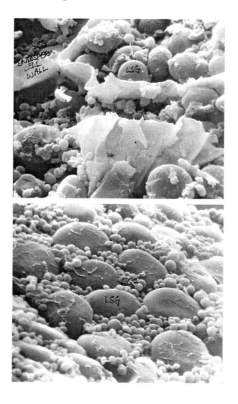

Fig. 58: Scanning Electron Micrographs (SEM) of barley cell walls and large (LSG) and small (SSG) starch granules. Fig. 59: SEM of well-modified malt after cell wall removal by malt enzymes.

MOISTURE Kilning reduces green malt moisture from 40–48% to 6% or less at a much faster rate than the moisture entered the grain in steeping. Equilibration of the residual moisture is thought to be the reason that two to three weeks of malt aging is necessary for most base malts prior to milling.

Raw grain moisture can and should be measured routinely before and after harvest, drying, storage, during malting, and in malt storage. Portable moisture meters costing a few hundred dollars are available (e.g. Agratronix, Alibaba, Dickey-John, GAC, Motomco, etc.) using near-infrared reflectance (NIR) spectroscopy, capacitance, or conductance technology to rapidly and accurately measure grains at 5 to 45% moisture.

There are also moisture balances that contain microwave, infrared, or halogen heat sources to actively drive off moisture in between initial and final weighing. These bench-top balances (e.g. Intell-lab, Ohaus) cost considerably more than portable units. Chemical standards of known moisture content (e.g. sodium tartrate) are available for drying standardization.

All rapid methods using portable or benchtop units must be routinely checked and calibrated against a standard Air-Oven reference method.

MUNICH MALT Munich malts are made in conventional kilns from moderate to high protein barley (11–13%), steeped to 44–47% moisture and germinated at 71–77°F (22–25°C). Kilning begins with a stewing period (100°F [38°C], and humid off-grain air recirculation) prior to drying at 120–150°F (50–65°C) and curing at 220°F (105°C) for light-colored and 250°F (120°C) for darker-colored Munich malts.

Munich malts are used to provide golden to reddish hues to beer, especially German *dunkels*, *marzënbiers*, and *Oktoberfestbiers*. Munich malts have a range of colors of 5–30 SRM, 10–60 EBC.

MYCOTOXINS Regardless of efforts to clean or sterilize grains, they always carry a complement of bacteria, molds, and yeasts, some of which can produce toxins that can interfere with seed germination and can be injurious to human and animal health.

Fungal mycelia occur in the field and in storage. Spores are found in grain dust, and can cause asthma, allergies, and "farmer's lung."

Pre-steeping in dilute solutions of sodium hydroxide, sodium hypochlorite, hydrogen peroxide, formaldehyde, lime, or sulfuric acid have been used to reduce the level of mold growth and mycotoxin contamination of malts.

NAKED OR HULLESS BARLEYS *(Hordeum vulgare distichon var. nudum)* Barley varieties with loose husk that fall off of the grain at harvest. Naked barleys have been used for food in parts of the Far East and Africa. Malting of naked barleys has been carried out, but care must be taken to protect the embryo, because the husk is not there as protection from physical damage and mold growth. Steeping times for hulless barleys, as well as all hulless grains (e.g. wheat) should be much shorter, often less than half as long, as grains with tightly adhering husks, otherwise over-steeping and drowning of the embryo can occur.

NATIONAL INSTITUTE OF AGRICULTURAL BOTANY (NIAB) A center for cereal crop research and plant breeding based in Cambridge, England founded in 1919 that makes annual approved crop variety recommendations, website www.niab.com.

NEAR-INFRARED REFLECTANCE (NIR) INSTRUMENTATION There are a plethora of companies that produce instruments that either reflect, "transflect," or transmit light onto small samples and measure certain wavelengths in the near-infrared spectrum (0.75–1.4 µm). These spectra are calibrated using provided software for determination of protein, moisture, oils, fatty acids, ethanol, and other constituents in whole grain, milled, powder, pellet, and liquid samples.

These instruments often yield analytical results in under 10 seconds, but they are expensive and are only as good as the accuracy, precision, and frequency of calibration to standard methods. Different grains, as well as different varieties of the same grain, often require unique calibration setups.

NITROSAMINES (N-NITROSO-DIMETHYL AMINE, NDMA) Nitrosamines are carcinogens formed when nitrogen oxides (NOx), formed by direct combustion of fuels, including peat, react with hordenine or dimethylamine on the surface of green malts. Air pollution, even in a micro-climate caused by an idling diesel truck near the malthouse fresh air intake, can result in higher NOx levels and greater risk of nitrosamine formation during kilning.

Burning sulfur-rich fuel oils or elemental sulfur in the kiln air intakes can mitigate nitrosamine formation in direct-fired kilns, however, formation of sulfur oxides increasing rates of corrosion to metal and concrete parts of these kilns.

Indirect kilning (the intermediate transfer of combustion heat to steam or thermal fluid) was voluntarily implemented in most malthouses immediately following a worldwide nitrosamine scare in malt, cheese, and beer in 1980. A corollary effect of indirect kilning actually increased malt fermentability by 2 or 3% (up to 88–90%) and spirit yield due to the absence of combustion gas compounds contacting the malt surface.

Nitrosamines are measured using gas chromatography and levels of <3.0 ppb are set as rejection limits in malt.

NON-CELLULOSIC ß-D-GLUCANS Most of the non-cellulosic ß-glucans found in plants, are "mixed-link" ß-glucans, which are polydisperse linear glucose polymers with approximately 30% of the links in (1,3) configuration and 70% (1,4) linkages. The high viscosity of mixed-link ß-glucans probably arises from extended linear shapes in solution.

NON-STARCH POLYSACCHARIDES Carbohydrate constituents of the cereal grain, not including the main polysaccharide, starch, can amount to 30–40% of total kernel weight and includes **ß-glucans, arabinoxylans**, and **cellulose**, which vary in their solubility characteristics and malting, brewing, and distilling importance.

NORTH DAKOTA STATE UNIVERSITY Department of Cereal Science—An NDSU sponsored program of barley and malt education for M.S. and Ph.D. students to gain knowledge and comprehension of the composition, functionality, and utilization of cereals or grains that will allow them to solve and analyze challenges within their field of employment.

Department of Plant Science—sciences of plant breeding and genetics, weed science, biotechnology, horticulture, agronomic crop production, turfgrass management, and cereal and food science.

OAT MALT Brewing with oat malt can add mouthfeel and toasted, biscuit aromas and flavors to beer. Oat malt yields about 75% extract, very low enzymes and high wort ß-glucan levels (300–400 mg/L). They can be kilned at low temperature for brewing in pale beers or roasted for darker colors.

Hubner et al. (2010) germinated oats for brewing at a range of times (48–144 hours) and temperatures (50–68°F, 10–20°C)and found that germination at 59°F (15°C) for 130 hours produced optimal levels of fermentable extract.

ORIGINAL EXTRACT (OE) Measured at the end of kettle boil, OE is the total extract from all brewing materials dissolved in mashing, lautering, sparging, and kettle boiling operations.

ORGANIC MALT Malted cereals that are grown, malted, and stored separately and labeled and sold as organic are regulated by the USDA's National Organic Program (NOP). Organic food and beverage sales represent about 5% of overall food and beverage sales in the U.S.

OVERGROWN KERNELS A percentage of green malt samples with **acrospires** extending beyond the distal tip of the grain, extending through the husk. Overgrown malts can lead to higher **malting loss** during germination due to increased acrospire growth and higher clean-off rates due to damaged husk.

P

PALE ALE OR MILD ALE MALT Malts made for production of traditional British pale ales with fairly low malt extract colors of 2–3 SRM (4–6 EBC). The British barley varieties Maris Otter, Halcyon, and Pipkin are traditional barleys for pale ale malts, but these high extract malts can be produced from any plump two-rowed malting barley with 9–10% protein levels.

PATENT MALTS The "Roasted Malt Act" enacted in the U.K. in 1842 stated that "malt is not to be roasted for sale, or sold, except by persons duly licensed." Roasted malts, as permitted by this license, or patent, were thereafter known by a term that is still used in some cases today, "patent malt." U.K. roasting laws also required that the roast house be separated from the malthouse by at least one mile.

Black patent malt, the darkest type of malt roasted in a drum, is used to brew dark porters, stouts, and black lagers.

Fig. 60: Malt storage crypt at Guiness St. James Gate, Dublin, Ireland.

MALT ROOM AT OLD BREWERY.

Fig. 61: Malt storage room at old Bass Brewery, Burton-on-Trent (Barnard, 1889).

PENTOSANS, OR ARABINOXYLANS A secondary group of structural polysaccharides found in cereal grains. The amount of pentosans has been reported as 2.3 to 4.0%, with the aleurone layer consisting of 85% pentosans. The quantity of pentosans is relatively unaffected during malting and shows an apparent increase during germination due to the degradation and loss of other grain constituents. However, the quality of pentosans remaining in malt has been linked to potential beer filtration problems. Pentosans and ß-glucans are lumped together in the terms gums (water soluble) and hemicellulose (alkali soluble).

PENTOSANASES *(arabinoxylanases)* Enzymes that can hydrolyze cereal pentosans include an arabinosidase, and endo-ß-1,3 xylanase, an exo-ß-1,3 xylosidase, and an exo-ß-1,4 xylosidase. Many of these enzymes are shown to originate from bacteria, wild yeasts, and filamentous fungi located on the outer layers of cereal grains. Recent research has focused on exploiting these microbes and their xylanases to assist with extract viscosity, lautering run-off, beer filtration, and flavor stability.

PERICARP A layer immediately underneath the husk that, joined with the **testa**, completely encapsulates the embryo, aleurone layer and endosperm. The pericarp and testa protect these cereal tissues in hulled and hulless grains from damage and soil microbes. Although water can slowly move into undamaged grains, larger molecules, like exogenous **gibberellic acid**, chemicals and even boiling sulfuric acid, cannot penetrate the thin but mighty intact pericarp/testa of grains.

PHENOLIC Any compound based on a ring of six carbon atoms joined by alternating single and double bonds. The tannins contained in grain husks are phenolic in nature, as are the soft hop resins (alpha and beta acids).

PHYSICAL TESTING OF CEREAL GRAINS PRIOR TO MALTING The ASBC Methods of Analysis (14th edition) Barley-2 lists a number of physical tests that can be performed on barley or any grain, including (A) Variety Determination, (B) Test Weight per Bushel (Bushelweight), (C) Assortment (kernel sizing using screens), (D) Thousand Kernel Weight (TKW or KW), (E) Texture of Endosperm (Mellowness), (F) Skinned and/or Broken Kernels, (G) Weathering and Kernel Damage, and (H) Injury by Sprout.

PHYSICAL TESTING OF CEREAL GRAINS DURING MALTING Physical methods of determining modification during malting are usually the simplest and most rapid tests to perform, and therefore, are most desirable to the practical maltster. These methods include grabbing a handful of grain at steep-out and counting percent **chitting** and routinely evaluating the aroma and flavor of germinating grains which can provide insights into plant hygiene, mold growth, grain health, and malt modification.

The **Maltster's Thumb**, described in detail above, is also a technique which can be done several times every day to determine the extent of endosperm modification.

PHYSICAL TESTING OF KILN-DRIED MALTS Friability as measured with a Friabilimeter or the **single kernel chew test** are physical method of evaluating dried malts.

Another physical test is known as the sinker/floater method and involves tossing 100 kernels of finished malt into a glass of water and stirring. Theoretically, grossly undermodified kernels will sink while highly-modified malts will float. Empirical experience is necessary for this archaic malt test.

Palmer (1999) describes the half-corn modification test, which involves shaking longitudinally-cut half corns of malt in hot water to wash out well modified areas of cut endosperms giving a visual indication of the variations in modification in individual grains which occur during malting. Some maltsters will hand-section their malts and use this method to assess **homogeneity** of modification.

PIECE, OR PILE The name given to a discrete unit or batch of grain as it moves through the malthouse. The malting piece or pile is the quantity of the grain, in pounds, as is, that is weighed out to begin the steeping process. Total malthouse capacity is calculated by the number of pieces produced per week, month or year times the starting grain weights of each piece.

PNEUMATIC MALTING Any malting system that uses fans or blowers to force air up or down through the grain during germination and kilning to control humidity and temperature and to remove respiration gases is called pneumatic malting.

In 1873, French maltsters Galland (of **drum malting** fame) and Saladin used ice to cool fan-driven air through grain in germination boxes with perforated steel floors. Immediate improvements in malt modification speed and uniformity were noted for this new technology.

Saladin later significantly improved pneumatic germination boxes by adding vertical helical turning screws allowing deeper bed depths, and avoiding the enormous labor involved in turning clumped, germinating grain by hand. Today, any rectangular, deep-bed (1 to 5 feet), perforated-deck germination boxes with mechanical turning machines are called Saladin boxes and have been built to contain from one to 1,000 tons of wet, steeped grain.

Turning machine helical screws whether purchased or manufactured, must have rubber or plastic wipers on the bottom that pull grains out of the screen holes, otherwise airflow will be reduced or even blocked by growing grain. Screws may be solid to lift grain up or made of horizontal bars to break through clumps or mats of grain horizontally. Spray bars should be mounted on the turning machine in case additional water is needed during turning to maintain or increase grain moisture.

Grain drying is minimized by ensuring that germination air is humidified at 95–100% relative humidity at all times, with no water droplet carryover onto the grain surface, which can cause excessive botanical growth and increased malting loss. When water is added through sprays or manually to prevent drying in pneumatic germination compartments a range of 40–80 gal/ton is often used.

Compared to traditional floor malting operations, pneumatic malthouses generally use much less labor for transferring grains from one step to the next and allow year-round malting with conditioned air.

> "Hail, Malt! Thou great Mother of Whiskey and Ale,
> Without thee how dull, and how poor each regale;
> Have I nought in my house save potatoes and salt
> And a big draught of Ale I'll ne'er be at fault."

POLYPHENOLS, ALSO CALLED TANNINS, PROANTHOCYANIDINS Polyphenols are complex aromatic compounds with two or more phenolic groups joined together. They provide astringency and mouthfeel, provide antioxidant properties and precipitate with soluble "haze-sensitive" proteins to make **chill** and **permanent hazes** in beer.

In an all-malt beer, roughly 70% of polyphenols come from malt and 30% from hops, aroma hops contributing more polyphenols than bittering hops, and winter barleys more than spring barleys. Higher pH and temperatures in steep water can reduce polyphenols in the mash, whereas these same conditions in mashing can increase solubilized polyphenols. Over-modification of malts can lead to higher polyphenols in beer because there are fewer high molecular weight proteins to precipitate with polyphenols.

Needless to say, balancing proteins from malts and polyphenols from malt and hops in mashing, boiling, "hot break" and "cold break" trub settling, and using fining and polyphenol-absorbing agents like polyvinylpolypyrrolidone (PVPP) and filtration are complicated aspects of brewing.

Commercial preparations of polyphenols extracted from oak tree galls are sold as tannic acid. These may be added when the polyphenol/protein ratio is out of balance requiring removal of proteins to avoid haze formation.

Low-proanthocyanidin and proanthocyanidin-free malts (e.g. ProAnt malt from Crisp Malting Co., Great Ryburgh, England) have been developed through traditional breeding methods which are claimed to improve colloidal stability of beers when used at 25–100% of the grist.

PRE-GERMINATION, ALSO CALLED SPROUTING Pre-germination can occur when grain embryos start to grow (produce roots and shoots), prematurely, while still in the ear. The antithesis of **dormancy**, which occurs in cool, damp weather, pre-germination occurs during periods of warm, damp weather before the mature grain is harvested. Pre-germinated kernels may be susceptible to damage and microbial infection during drying and storage and will malt unevenly. Most grain specifications set maximum allowable pre-germinated kernels at <5%.

The ASBC method for quantifying pre-germination is Injury by Sprout and consists of removing the husk with a mechanical pearling machine and then visually counting sprouted grains in the sample.

The **Falling Number method** is used by maltsters and bakers to assess flour quality with respect to pre-germination. Incubation of ground flour with Phadebas reagent (dye-linked amylopectin), or staining half-kernels with fluourescein dibutyrate or tetrazolium chloride can also be used to measure pre-germination.

PROTEIN (NITROGEN), SOLUBLE, TOTAL, AND SOLUBLE/TOTAL PROTEIN RATIO Proteins are multi-functional components of cereal grains, accounting for a large percentage (50% or more) of plant cell weight. In raw grains, proteins provide the structural matrix for the deposition and protection of carbohydrate (starch granules and sugar molecules). During cereal germination, some proteins have an active role as enzymes to solubilize (wet) and degrade (remove) cell wall and protein matrix materials.

Usually about half of the total proteins in barley are hydrolyzed to make them warm water soluble (%Soluble/Total protein ratio of 40–45%). These soluble proteins provide enzymic activity and **free-amino nitrogen (FAN)** in the wort and are critical to adequate sugar formation and yeast nutrition.

After malting is complete, and during brewhouse or distillery mashing processes, proteins provide low molecular weight peptides and 19 amino acids essential for yeast nutrition during fermentation. Residual proteins in beer contribute to its mouthfeel, body, head retention, and can cause temporary and permanent hazes, if not properly controlled by gravity in brewhouse hot and cold breaks and cold beer settling or removed later by mechanical filtration.

Breeding programs and grain contracts typically set maximum protein levels around 13.5% in North America and 11.5% in Europe.

Soluble and total protein content is measured using chemical (**Kjeldahl** or **Dumas**) methods or scanned using near-infrared reflectance (NIR) instruments which have to be calibrated routinely against chemical methods. Protein content is estimated by multiplying measured nitrogen content by an averaged factor of 6.25.

In the 1930s, L.R. Bishop developed the Bishop Regularity Principle, which simply states that proximal analysis of cereal grains is a zero-sum game: for every 1% increase in barley protein, there is a 0.5 to 0.7% decrease in potential malt extract from that barley. This principle also applies to other malting grains, albeit with unique replacement factors for each.

Total barley protein levels of 10–12% are considered ideal for most malting applications, although the Brewers Association (2014) has recently set 10.5% as a maximum desired level. Grain purchase contracts usually specify purchase and premium criteria for protein levels meaning that grain growers must be careful with their fertilizer applications.

The lowest level of protein suitable for malting, brewing, and distilling is not absolutely known, but total protein of 9% or less may fall short in production of enzyme activity, amino acids, and FAN for yeast nutrition. Historically, prior to wide-spread nitrogen fertilizer application, malting barleys were harvested at very low (7–9%) protein levels, but the very long malting times (2–3 weeks) produced sufficient FAN to support all-malt fermentations.

Total protein in raw and malted grains can be analyzed by **Near-infrared reflectance** analyzers. Soluble proteins and FAN can be analyzed by combustion methods, colorimetric methods, or Auto-analyzer.

Soluble malt proteins are composed of large polypeptides (involved in beer foam and haze); smaller peptides (2–10 amino acids in length) that contribute to yeast nutrition and beer mouthfeel and body; and even smaller amino acids, representing 10–15% of extract soluble protein, that are essential for yeast nutrition during fermentation.

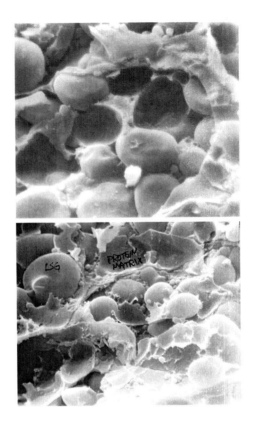

Figs. 62 and 63: Barley protein matrix and small and large starch granules.

PROTEIN REST, PROTEIN STAND An early mash hold temperature, usually 110–130°F (45–55°C) sometimes used in brewhouses to protect heat-labile proteolytic activity in order to increase **FAN** content in worts prior to fermentation. It is believed that 20–25% of total wort FAN comes from the early stages of mashing.

PROTEOMICS The fairly new discipline of proteomics is the large-scale study of proteins by mass spectrometry to detect (with high specificity) glutenins, gliadins, and related prolamins in flour, food, and beer. Recent research has revealed that protein structural changes and post-translational modification can modulate the flavour, texture, and appearance of beverages (Colgrave et al., 2013). These authors claim that proteomics, using gel-based spectrometric and mass spectrometry techniques will help future maltsters and brewers—in association with breeders—create better and safer products for consumers, and characterize novel proteins.

PTYALIN The name given to α-amylase found in human saliva that has the same starch-degrading functionality of malt α-amylase. Some native African and Latin American beers (e.g. chicha) actually depend on salivary enzyme hydrolysis of manioc and other starches that are chewed and spit into communal vessels prior to fermentation.

QUARTER An archaic term used in the U.K. and South Africa until the early 1970s for measuring barley and malt. One quarter of barley (8 bushels, 448 lbs.) would produce a quarter of malt (8 bushels, 336 lbs). "Sacks" of barley and malt were each ½ quarter, or four bushels.

QUINOA *(Chenopodium quinoa or C. nuttalliae)* Quinoa is a very small pseudo-cereal which makes it gluten-free. It contains bitter-tasting saponins on its surface, which have some antibacterial properties. Its small size means quinoa absorbs water very rapidly, requiring short steeps. Quinoa can be malted as follows: Steeping, 59°F (15°C) with water and aeration steps, 3W–3A–3W–steepout at 44–48% moisture; germinate at 59°F (15°C) for 5 days; kiln at 122°F (50°C) for 16 hours, 140°F (60°C) for 1 hour, 149°F (65°C) for 5 hours.

RACHILLA The basal bristle located where the barley grain is attached to the plant stem. The rachilla lies in the central dorsal groove of the kernel and is used (long and short rachilla) to help in the identification of barley varieties by examining single kernels.

REAL EXTRACT (RE) Because ethanol is less dense than water, after creation by fermentation, it can interfere with some extract density measurements. Real extract is calculated by the removal of ethanol by distillation and volume and temperature corrections applied. Real extract can be back-calculated after fermentation from the initial wort, °Pi; and fermented beer, °Pf; gravities (degrees Plato) are known:

RE = 0.1808 (°Pi) + 0.8192 (°Pf)

REGULATIONS TO BUILD AND OPERATE A MALTHOUSE In the U.S., specific zoning rules and regulations for food plants vary by state, county, and local municipality. These regulations may include building and fire codes and certificate of occupancy (CO), food manufacturing and agricultural licensing, Food and Drug Administration (FDA) registration for livestock feeding of spent grain, wastewater utility quantity and quality approvals, sales and use tax certification, construction fire regulations and routine inspections, fugitive emissions (dust and odors), noise, alarm permits for security systems, and others. Check with all local and state municipalities before you begin construction.

All of these hurdles may seem daunting, but they pale in comparison to what British maltsters had to comply with in the 19th century. In addition to Royal regulations regarding the construction of steep cisterns and number of germination working floors, and documentation in barley and malt books, floor maltsters of the day were required to provide written notice in advance (24 hours for in-town maltings and 48 hours for those in the country) of the exact time and duration of steeping and the quantity of grain to be wetted (Clark, 1998), with penalties for failing to meet any one of the 14 steeping regulations of £100–200, ($10,000–20,000 in today's money!).

Reinheitsgebot, **OR GERMAN BEER PURITY LAW** The German beer purity law dates back to 1516, when Bavarian Duke Wilhelm IV declared that commercial brewers may only use water, malt, and hops to make beer. Yeast was unknown at that time and so was not specifically named until the 17th century. This law was originally called "*Surrogatverbot*," meaning surrogate or adjunct brewing materials were forbidden. The term "*Reinheitsgebot*" was not applied until 1918 by the Bavarian State Parliament. Malt is defined as any grain that has been caused to germinate in a malthouse.

RESPIRATION OF GRAIN DURING STORAGE, STEEPING, AND GERMINATION

During cereal grain germination in the field or in the malthouse, respiration is the source of energy for all of the plant's synthetic processes. Early in germination, respiratory activity is evenly distributed between the two respiratory tissues, the embryo and aleurone layer. Later in germination, respiration occurs mainly in the embryo, peaking at three or four days. Respiration converts grain carbohydrates into carbon dioxide and generates heat, resulting in ~4% loss of weight. This production of CO2 and buildup of heat of respiration are the reasons extraction fans are needed in steeping and germination cycles in pneumatic malthouses.

As long as oxygen is available to the embryo and aleurone layer, cereal grains will respire, producing carbon dioxide, heat, and water vapor, from glucose hydrolyzed from starch according to the following formula:

$$C_6H_{12}O_6 + 6\ O_2 \rightarrow 6\ CO_2 + 6\ H_2O \text{ and } 674 \text{ kcal energy}$$

If oxygen is lacking for extended periods during grain storage, steeping or germination, intermolecular respiration forms alkanals and alkanols, which can poison the embryo, impeding germination and normal malt modification.

The amount of respiration during raw grain storage as measured by CO2 generation is highly influenced by grain moisture (Kunze, 1999). Grain at 11% moisture generates approximately 0.3 mg CO2 per kg barley in 24 hours storage at 68°F (20°C), whereas the same grain at 17% moisture will produce 300 times that much CO2 (100 mg) in the same time period, leading to elevated pre-malting respiration losses and risks of mold growth and embryo cell death.

Higher grain storage temperatures will also increase the amount of CO_2 evolved from respiration. One kilogram of barley stored at 59°F (15°C) will generate 1.4 mg of CO_2 in 24 hours, and will produce nearly 200 times that much CO_2 (plus heat and water vapor) when stored at 126°F (52°C). For this reason, it is always advisable to draw cold, dry air through grain storage bins or flat storage at every opportunity.

Cereal grains with smaller kernels, or those without protective husk layers (e.g. wheat and hulless barley) are even more at risk of respiration damage during storage, resulting in safe storage time periods about 30% shorter than normal, husked barley.

RICE At least one U.S. Craft Maltster (Eckert Malting and Brewing, Chico, CA) is malting rice and kilning to produce a range of colors from pale to black.

Fermented rice foods in Asia and Africa include *bhattejaar, brem, busa, chongjiu, doburoku, khaomak, laojiu, makkori, mirin, sake, satou, shaoshingjiu, tape, tapuy, thuon*, and *zutho*.

Fig. 64 (left): Jim Eckert and his rice malt roaster. Fig. 65 (right): Rotating basket micro-malting machine by Alloway Standard Industries Fargo, ND at Eckert Malting and Brewing Co. Chico, CA.

ROUSING During wet immersion cycles in steeping, compressed air can be introduced from lines and nozzles distributed across the bottom of the conical or flat-bottomed steep tank, which serves to oxygenate, cool, and stir the grain during steeping. Aeration rate during rousing should be 2.0–2.2 ft3/min/ton.

RYE MALT Extract of malted rye can be very high (85–87%). The color value of rye malt extract is higher than that of malted barley (7–12 SRM [14–24 EBC]) and can add a reddish hue to beer or whiskey. Rye flavor notes can be very dominant in beer. A little rye can add toffee notes to beer, whereas higher levels can add a very spicy, lingering after-flavor.

S

SALADIN BOX GERMINATION In the 19th century, French engineer Charles Saladin developed concrete boxes with rotating helical auger turning machines for germination in pneumatic malthouses. Dimensions of Saladin boxes can be 7-to-1 (length/width), up to 150 feet long, have slotted floors, and nowadays made of stainless steel. Forced airflow in Saladin boxes can either be up- or down-draft.

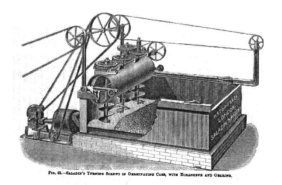

Fig. 66: Saladin turning machine (Stopes, 1885).

Fig. 67: Saladin box turning machines at Colorado Malting Co. Alamosa, CO.

SAMPLING Obtaining representative samples from grain bags, bins, or tanks is a very difficult and often over-looked aspect of analysis. Sample triers, spears, diverters, and vacuum devices (both manual and automatic) are available to probe and improve accurate retrieval of samples from different depths in bins, trucks, and silos.

Milled flour is the most troublesome type of sample to collect because particle sizes, densities, weights, and even static charge can interfere with usual sampling efforts.

Prior to paying for expensive laboratory analyses, the grain grower and maltster must ensure that samples are completely representative of the entire batch.

SCALD *(Rhynchosporium secalis)* A barley field fungus that can cause blotches with dark brown edges surrounding bleached leaf lesions that can kill a leaf and infect heads and grain. Water splash from rain or irrigation can spread the disease. Resistant varieties and foliar fungicides are available.

SEMI-CONTINUOUS MALTING Moving grains automatically and continuously from the bottom to top of a tank, or between malting floors or "streets" is called semi-continuous malting. Examples of these in the past have included the Engerth system, developed in the late 19th century, in which grains moved from the bottom of germination tanks to the top with belts and elevators; the Lausmann Transpose system, which used a continuous cupped belt system to scoop grain from the top of a series of tanks as the perforated decks are hydraulically raised or lowered; and the Wanderhaufen or "walking pile" continuous germination "street" system with moving perforated deck and inclined turning machines that lifted and dropped grains farther along the moving bed until, at the end of the street, green malt is dropped to a moving deck kiln below.

SLACK MALT Kiln-dried malts are hygroscopic, meaning they will absorb moisture from ambient air. Because warmer air can hold more water vapor than cooler air, malt should be kept as cool and dry as possible during the entire storage period. Slack malts can exhibit mash filtration run-off difficulties, reduced extract and fermentability, as well as off-flavors from oxidation of fatty acids.

SMASH BREWING An acronym for Single Malt and Single Hop beers that are gaining recent attention as means for brewers to evaluate and compare flavor and efficiencies of two different malts or two different hops in the same recipe.

SMOKED MALT, *or Rauschmalz* The first malt kilns were either sun-dried or wood-fired with direct contact of wet, green malt with wood smoke, picking up varying degrees of smoke aroma and flavor in the final product. Some beers, like German Rauschbiers (smoked beers), are still made with definite smokey flavors today. The German word for "smoked malt" is *Rauschmalz*, originating in the town of Bamberg. Rauschmalz is traditionally kilned over an open, smoldering fire of peat or woods like beechwood, alder, almond, apple, or cherry wood.

Of course, smoke from dried peat fuels have been used in Scotch whisky production for hundreds of years. Smoked malts, therefore, have heretofore been defined in fairly narrow interpretations in making malt for beer and whiskey until very recently (Bell, 2011).

Perhaps the most innovative malt smoker in the world of whiskey production today is Darek Bell, owner/distiller of Corsair Artisan Distillery (Nashville, TN) who reports that he has smoked malts and made whiskies with many different varieties of wood and bark, including: alder, almond, apple, apricot, ash, avocado, bayberry, beech, birch, blackberry, black walnut, butternut, cabernet barrel wood, cascara, catuaba, crabapple, cherry, cedar, cottonwood, dogwood, fringe tree, grape, Hawaiian guava, Hawaiian kiawe, Hawaiian ohia, hickory, Jamaican dogwood, lemon, lilac, macadamia, manzanita, maplewood, mesquite, muira puama, mulberry, nectarine, olivewood, orange, pimento wood, prickly ash, peach, peat, pear, pecan, persimmon, plum, quassia, red oak, sassafras and white oak, white willow, and yohimbe.

Spices and herbs that Corsair have also burned in experimenting with unique smoked malt aromas and flavors include: angelica, anise, calamus root, catnip, clove, ginger, jasmine, lavender, lemon balm, licorice, mint, mugwort, osha, peppermint, sage, skullcap, spearmint, sweet cicely, terragon, violet, wormwood, and yerba buena.

The book on smoked malt is much wider-ranging than the traditional uses and is currently being re-written by entrepreneurs like Bell.

SMUT A disease of barley that can cover entire plants with sporing fungal masses (loose smut and covered smut), which will kill the embryo and wipe out entire crops. Proper crop rotation and maintaining clean seed are used to control smut from year to year.

SORGHUM *(S. vulgare or S. bicolor)* Sorghum is the fifth most widely grown cereal crop, after wheat, maize, rice, and barley. Malting of sorghum is done routinely in Africa, and is being evaluated worldwide as a potential **gluten-free malt**.

Sorghum grains can be white, brown, purple, or red, owing to its high content of **polyphenols (tannins)**. Pre-washing sorghum grains with formaldehyde or sodium hypochlorite (NaOCl, bleach) has been done to reduce tannin content, followed by clear water for steeping. Both of these chemicals, however, can be poisonous and must be rinsed off completely.

Sorghum malt is made by Thomas Fawcett and Sons Malting Company (Castelford, West Yorkshire, U.K.), which is used to produce a supermarket light beer described as gluten-free. Like most traditional sorghum beers, it is sour. Sorghum is also used commercially as an adjunct to produce standard beers.

Red, actively fermenting sorghum beers are still brewed in parts of Africa and are variously called *bouza, burukutu, dolo, ikigage, kaffir, kishk, muratina, pito, seketeh, shakuparo, bogobe, injera, kisra, koko,* and *ogi*.

Sorghum can be very difficult to malt, requiring high steep-out moisture (50–55%) and very warm germination temperatures of 75–85°F (25–30°C). Because sorghum has no protective husk, it is more susceptible to entangled rootlets, grain damage, and mold growth. Sorghum becomes so matted during germination that it must be turned during kilning.

Adequate sorghum malt quality has been achieved by steeping at 81°F (27°C) in a wet immersion (W) and air-rest (A) steep cycle (with hours of each step) of 10W–4A–10W–7A–15W–4A–steepout; germination for 6 days at 81°F (27°C); and kilning at 122°F (50°C) for 24 hours.

Mycotoxin contamination is of concern when malting with sorghum, leading to recent research into adding formaldehyde, dilute alkaline solution, or adding solutions of *Bacillus subtilis* as biocontrols to compete with mold growth (Tawabe et al., 2012).

The Kimberley process is a production of native African sorghum beers using simultaneous sorghum-lactic acid fermentation and a milled sorghum and maize grits fermentation, followed by fermentation with dried brewer's yeast. Like most sorghum beers, it is quite sour, turbid. and meant to be consumed while actively fermenting.

SPACE BARLEY Japanese brewer Sapporo introduced a beer named Space Barley in 2009, brewed with malt made from fourth-generation descendants of Haruna Nijo barley seeds that spent five months in space at the International Space Station in 2006.

SPECIALTY MALTS A term used to describe malts other than the standard brewers' malts that are added in very small quantities to impart flavor and sometimes color to the final product. Brewers and distillers consider most specialty malts as adjuncts in extract, enzyme, and fermentability calculations, rather than workhorse base malts, because stewing or roasting in the kiln usually inactivates all or most malt enzyme activity in these malts.

Specialty malts, described in more detail following, can be produced from raw, chitted, germinated, or kiln-dried grains and include **amber malt, black barley, black malt, brown malt, caramel malt, crystal malt, chocolate malt, dextrin malt, Munich**, and **vienna malt.**

A wide range of colors and flavors can be produced when roasting malts in an externally heated rotating drum kiln. Unlike pale malts, which require uniform distribution of kernel moisture facilitated by malt aging after kilning for optimum extraction and fermentation, most roasted malts do not require any storage prior to brewing or distilling, other than time waiting for malt analysis to be completed. In fact, roasted malts can actually lose some of their fresh, volatile aromas and flavors and become oxidized if stored for too long a time period after roasting and before use.

SPELT *(Triticum spelta)* An heirloom hard-grained, hulled wheat variety, which is also called *Dinkel* in Germany. Malted spelt has very high enzyme activities and low color value of 2–4 SRM (4–8 EBC) and can replace nearly 50% of malted barley in brewing or distilling grist, producing acidic, nutty flavors in the beer or "alt" whiskey.

SPENT GRAINS, OR DRAFF After liquid, sweet extract has been separated by mash filters, lauter tuns, Strainmasters, or cloth-draped plate and frame horizontal mash filter, the insoluble solids (husk, protein, lipids, and some sugary wort) are called spent grains.

Moisture content of spent grains is very high, around 50–60% from mash filters and 75–85% from lauter tuns, therefore it can weigh 20–40% more than the original malt bill.

Dried spent grains can contain 28% protein, 40% carbohydrates (cellulose and ß-glucans), 10% fat, and 6% minerals.

Wet, or dried and pelletized spent grains, are usually sold or given away as animal feed, compost fodder, fish food, mushroom compost, etc. Applications of spent grain in baking, tortilla chips, concrete mixes, and biogas generators have also been developed commercially.

Necessity being the mother of invention, Alaskan Brewing Co. in Juneau, AK—with no roads in or out—had been drying and shipping their spent grain to farmers and ranchers in the Pacific Northwest for 20 years. In 2011, they developed a first-of-its-kind steam boiler fueled entirely by spent grain. Since then they have reduced boiler fuel-oil use by more than 65%.

SPIRIT YIELD The amount of pure distilled alcohol (liters) that is obtained from one metric ton of malt (dry weight) as measured by distillers.

SPRING BARLEY Spring barleys are planted in the spring and harvested four to five months later in late summer. If, like **winter barley**, spring barley is planted late in the fall, it will freeze and die. Most malting barleys are spring barleys.

SQUEEZE MALTING A process developed in the early 1980s designed to minimize the amount of water, and therefore energy, used to de-water malt in kilning. Grains are steeped to about 36% moisture then squeezed between rollers with a gap size of 1.8 mm, which partially dewaters and damages the grain, similar to **abrasion systems**, so that exogenous **gibberellic acid** (GA) can be applied for faster malt modification.

STARCH The most abundant reserve carbohydrate of higher plants is starch, encompassing 20–70% of all undried plant material and 55–65% of barley by weight. The word "starch" comes from the Anglo-Saxon stearc meaning "strength" or "stiffness."

Cereal starches are a mixture of two polysaccharides, **amylose** and **amylopectin**, both of which are polymers of glucose (α-D-glucopyranose) and named for Amylon, the Greek word for starch.

Barley endosperm contains starch that is packaged during grain development in amyloplasts, resulting in an essentially bimodal distribution of starch granules.

Small granules, which are approximately 3 μm in diameter, constitute ~90% by number but only 10% by weight of the starch granules in barley. Large starch granules, ~30 μm in diameter, make up the balance.

Alpha-amylase is the only enzyme that can attack native starch granules, producing granule pitting which is easily visualized microscopically in hand-sectioned malt kernels. Over-modification in areas of sectioned malt can be observed by the disappearance of small and pitting of large starch granules.

The carbohydrases that hydrolyze starch are referred to collectively as amylolytic enzymes.

STARCH CONTENT OF CEREALS, FROM SHELTON AND LEE (2000).

CEREAL	STARCH, %
Barley	57–60
Brown Rice	64–68
Corn (maize)	64–78
Milled Rice	75–79
Millet	61–65
Oats	43–61
Rye	65–71
Sorghum	60–77

| Triticale | 50–56 |
| Wheat | 63–72 |

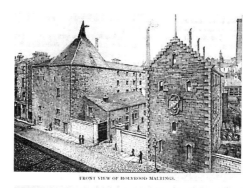

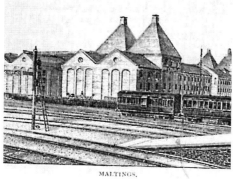

Fig. 68 (above): Holyrood Maltings, Edinburgh, Scotland. Fig. 69 (below): Ind Coope Maltings, Burton-on-Trent, England (Barnard, 1889).

STEEPING Steeping is the first and most important of three main process steps in malting. Its purpose is to provide additional seed cleaning, initiate uniform germination of the grain, and hydrate the cell walls and proteins of the endosperm to get them ready for enzymic modification in **germination**.

Simply stated, steeping must allow 100% of the grains

to germinate, because there is no practical way to remove un-germinated (dead) grains that will have highly disproportionate negative effects on malt and beer quality. All kernels should absorb water at the same rate. Kernels with excessive moisture will either die or germinate too rapidly, and low moisture kernels will not allow efficient water-mediated (hydrolysis) modification in time. Once germination begins, all grains must grow at the same rate to produce a finished malt of optimal and uniform modification.

An often over-looked area of steeping is the final chance to remove dust, field dirt, and husk tannin (polyphenol) by water soaking and rinsing.

The hydration of cereal grains in the malting process is facilitated by a series of soaking, air rests, and spray cycles, which constitute the steeping process. Hertrich (2013) elegantly described the maltster's management of immersions and air rests in steeping:

> "Maltsters consider air rests as the accelerator of the steep process and immersions as the brakes. It is more about changing the respiration rate and the need for the barley to "drink" from the next immersion than about the length of the next immersion."

Modern steeping methods require 24–80 hours; whereas, 300 years ago, winter steeping periods of 150–174 hours were not uncommon (Tryon, 1691). Regarding the quality of water used, Tryon recommended river water to make "mault" because of the "…benevolent Influences of the Celestial Bodies running in the open Air and Light of Heaven." The water used in malting today is more precisely defined (Moll, 1979) as to its physical, chemical, and bacteriological properties (see **Water**).

As described by Sir Robert Moray in 1678:

> "Take then good barley newly threshed, and well purged from the chaff, and put eight bolls of it, that is, about six English quarters, into a stone trough; where let it infuse till the water be of a bright reddish colour, which will be in about three days, more or fewer, according to the moistness or dryness,

smallness or bigness of the grain, and according to the season of the year or temper of the weather; in summer malt never makes well; in winter it will require a longer infusion than in spring or autumn."

At the start of steeping, the initial uptake of water causes the entire grain to swell. This swelling, mainly due to the imbibition by protein, which constitutes 9–14% of the grain, increases the grain volume by ~25% and total weight by 50% (Hough et al., 1971) and must be taken into account in steeping, germination, and kilning vessel and conveyance equipment designs. Imbibition by the cell wall material may also contribute to swelling, but the starch does not swell under normal steeping conditions.

Final out-of-steep moisture content desired for optimum malt modification, with minimum grain damage and malting loss, has traditionally been 42–44% for pale ale, lager, and pilsner malts, and 44–47% for darker ale and distilling malts, but can vary with cereal grain characteristics and malting process.

Hartong and Kretschmer (1958) measured the rate of water uptake by barley during steeping and found a positive correlation to maltability. The same authors (1969) reported the use of water uptake as a purchasing criterion for barley and compared the results to malt quality.

The functions of water, once absorbed into the grain, include cell elongation, respiration, and secretory activity of the embryo (production and secretion of gibberellic acid), activation of enzymes, which begins the hydrolysis and mobilization of the cereal seed reserves and hydration of the endosperm cell wall for subsequent hydrolysis by released and synthesized enzymes.

An old floor maltster's adage that says "steeping hours plus the temperature in degrees Fahrenheit should equal 110" is a good starting point when steep water temperature is variable or a new cereal grain is being evaluated. For example, steeping with 50°F water might take as long as 60 hours, whereas using 70°F water should take ~40 hours. Water uptake during barley steeping has been shown to occur in three stages (Brookes et al., 1976).

In the first stage, physical adsorption raises grain moisture to 32–35%, including the husk, which swells to 50% moisture rapidly, and the embryo to 60% moisture. The rate of water uptake is directly proportional to water temperature, taking just a few hours at 77°F (25°C) and up to 12 hours at 50°F (10°C).

Dissolved oxygen levels in the first steep soak water drop precipitously to zero in the first hour of soak. Replacing some of this dissolved oxygen and stripping out evolved carbon dioxide is achieved by continuous bubbling or **rousing** during soak cycles and downdraft CO_2 evacuation during air-rest cycles. Some brewers specify placement, size, and airflow requirements of air nozzles on steep tank cones and walls of commission malthouses. Some also require a certain continuous overflow of fresh water during steeping soak cycles, although this requirement is being re-evaluated with respect to water conservation. If not stripped out or reduced, CO_2 and small amounts of ethanol produced during steeping can inhibit grain growth.

Other compounds released during steep soaks include polyphenols, amino acids, sugars, and salts from the husk, which are removed during steep soak water drains.

The second stage is a lag phase, lasting for about 10 hours, during which time little water is absorbed, therefore steep cycles normally go into an air-rest/CO_2 evacuation step of 12–16 hours during this stage.

In the third stage, water uptake rate increases as cells elongate and begin chitting. Embryo moisture rises to a maximum of about 85% after chitting. The endosperm is the last tissue to hydrate during steeping, with areas of higher protein and starch densities requiring the maltster's patience to ensure proper hydration for modification to occur.

Vigorous aeration during steep soaks, spray steeping, high temperatures, or poor quality barley can upset the normal balance of water within the various tissues of the grain. If temperatures and soak/air rest cycles are not carefully controlled, the rapidly growing embryo and aleurone tissues may actually syphon water away from each other or the endosperm, interrupting enzymic hydrolysis and uniform malt modification. For these reasons, cooler, longer steep cycles are generally preferred over shorter, warm steeps.

Steeping traditionally was broken into three or four immersions, interspersed with short periods of air rest, or CO_2 evacuation. In order to save water, two immersions are sometimes used, with additional water added via sprays on the turning machines in germination. Water usage and wastewater treatment are reduced, but excessive root growth may occur if water is sprayed after rootlet emergence (chitting), leading to uneven and longer germination periods and higher malting loss.

Once steeping conditions have been established for a grain based on laboratory GE, GC, and WS and pilot or "tea-bag" malting trials, the steeping recipe should specify:
- Number and cycle times of soaks and air-rests
- Grain:water ratio during each soak
- Immersion grain–water slurry temperature range, e.g. ±4°F (2°C).
- Time to level grain during each soak, e.g. within 15 minutes.
- Soak times hours and periods of rousing (mixing)
- Air-rest times and CO_2 extraction fan check
- Moisture target at end of each air-rest period (after surface water is absorbed), e.g. for two-rowed barley: first air-rest, 36–40%; second air-rest, 42–45%.
- Temperature difference maximum during air-rest periods (top-bottom and side-side in tank), e.g. ±5°F (3°C).
- Steep-out moisture (after final soak and transfer water absorbed), e.g. for two-rowed barley, 45–48%.

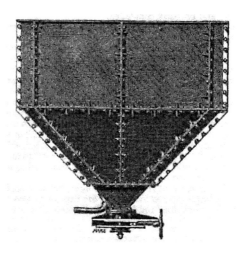

Fig. 70 (above): Stopes steeping istern (Stopes, 1885). Fig. 71 (below): Micromalting "tea bag" of Maris Otter barley at Tuckers Maltings UK. A scoop for removing floaters from the 100-year old steep cistern is in background. (photo courtesy of Roger Putnam).

Fig. 72: Malt chariot at Tucker's Malting, U.K. (photo courtesy of Andrea Stanley).

STEEPING DEGREE DETERMINATION FROM WAHL AND HENIUS (1908)

- When cutting through a grain, the contents should show completely and uniformly wetted, with the exception of a minute speck in the center of the endosperm.
- When taken by the ends between thumb and index finger, and pressed, the kernel should not prick the skin.
- The kernel should be elastic enough to be bent over the fingernail without breaking.
- At the end where the radicle is located the hull should appear to be open.
- Upon biting gently into the kernel, the endosperm should move to both sides without breaking or cracking.
- A sample of six-rowed barley, f.i. Manchurian, taken from the steeping tank should show and increase in weight of about 60%.

- Of these indications, Nos. 1 and 6 are the most reliable.

STEEPING TECHNOLOGY Steeping vessels may be constructed as conical, flat- (or false-) bottomed tanks, steeping drums or vertical steeps made from stainless steel or plastic. Each of these vessel designs have advantages and disadvantages of construction and operation.

Fig. 73 (above): Floor malt at steep-out (casting). (Photo courtesy of Andrea Stanley) Fig. 74 (below): New York Craft Malt LLC (Batavia, New York) 1-ton germination and kilning vessel (GKV).

Conical steep tanks—The simplest stand-alone steep tank design consists of a cylindrical tank with an open conical bottom. The conical bottom in these tanks is full of grain and water (during soak cycles) and grain only (during air rests). Conical steep cone angles facilitate the gravity flow of grain and water during steep-out with little manual intervention. The conical tanks can make it difficult, however, to get even distribution of rousing air through compressed air nozzles positioned all around the cone, which is intended to push air up through the grain for leveling and cooling purposes, and CO_2 evacuation air, pulled down through the grain during dry periods, or air rests.

If the diameter of the conical steep tank is too small relative to grain height in the tank then grain swelling has been observed to produce a compacted ring of grain near the tank periphery. This compaction interferes with water and air movement through grain near the outer tank circumference, resulting in uneven germination.

An improvement to conical steep tanks uses a central pipe that is open at the top and bottom and fitted with a compressed air injection nozzle near the bottom to get the grain and water to roll up through the central pipe and down the outside tank walls during rousing, improving circulation and grain washing and preventing grain compaction.

Double-walled stainless steel cylindro-conical steep tanks are further improved by installation of perforated decking across the entire area of the tank located at the junction between the upper cylindrical and lower conical tank parts. There are pneumatically driven steep-out gates that open during steep-out to send water and grain down to germination compartments facilitated by the cone angle. The perforated-deck plates have the advantage of ensuring even distribution of rousing air during wet soaks and evacuation flow during air rests, but require more water during steep soaks because the conical plenum below the grain-holding perforated deck, although containing no grain, must be filled with water during each water soak cycle.

Flat-bottomed (or false-bottomed) steep tanks—These cylindrical tanks are constructed with a slotted stainless steel plate above a flat floor of the tank, rather than the conical tanks as in the MillerCoors malthouse. Compressed air pipes and nozzles are located below the plenum.

A new type of false-bottomed "ECO-steep" vessel, with

a diameter of 69 ft (21 m), has recently been commissioned at Bairds malthouse in Arbroath, Scotland (Kinsman, 2011). The under-floor plenum has been replaced with a network of pipes, valves, and pumps below the slotted steep floor which reduces water usage and still provides proper aeration and CO_2 extraction distributed over the entire steep tank.

Separate vessels for Steeping (SV), Germination (GV) and Kilning (KV)—This configuration of separate vessels is the most common type of malthouse, providing the greatest flexibility and optimization of production output as well as improved opportunities for plant hygiene. Initial capital and maintenance costs for the extra valves, pumps, conveyors, and elevators needed may be slightly higher, as can labor costs.

Steeping and Germination vessels (SGV) with separate kilns—Mainly due to considerable swelling of cereal grains in steeping, requiring grain movement to break up compacted spots prior to forcing air through it in germination and kilning, this option is not recommended or common.

Germination/Kilning Vessels (GKV) with separate steeps—This was the most popular option for new malthouses constructed in the 1960s and 1970s, also called Fleximalt systems. Chitted malt from separate steeping vessels is transferred to pneumatic GKV boxes for the remainder of the germination and kilning processes.

There are three advantages to this combined malthouse process, first is that swollen grains from steeping are transferred either hydraulically (wet steep-out) or via manual methods or machinery (dry steep-out), ensuring breakup of any compacted spots caused by grain swelling during steeping. Second, malthouse capacity can be increased 15–25% by starting the next steep-in a day or two before end of germination and kilning. Third, though requiring more energy than separate germination and kilning vessels, the act of heating the GKV every 4–6 days to kiln the grain helps to control mold growth, which can be a problem in separate germination compartments (GV) that are perennially cool and humid.

The major disadvantage to GKVs is that any ridges or furrows made by turning machines during germination will be left on the grain surface going into the kiln cycle, creating air channeling and uneven grain drying.

Around 120–140 lb/ft2 (original barley weight) was found to be acceptable grain loading for combination Germination-Kiln (GKV) vessels (Gibbons, 1989).

Fig. 75: Four-ton capacity steep-germinate-kiln vessel (SGKV) manufactured at Valley Malt (Hadley, Massachusetts).

Steep-Germinate-Kiln Vessels (SGKV)—Single vessel malthouses that combine all three steps in one compartment, tank, or drum are the most efficient regarding floor space (except for tower malthouses), although they also have drawbacks. Like Steeping/Germination Vessels (SGV), the lack of complete breakup of compacted zones in steeping may lead to air channeling, grain souring, and wet spots in germination and kilning. Also, total plant throughput is limited and vessel turnaround times extended, because prior kiln runs must be completed, emptied, and cleaned before starting the next steep in the same tank.

Gibbons (1989) compared SGKV malthouses in the U.K. and found that SGKV plants had higher **capital construction costs** and higher operating costs than those with separate process vessels.

Gibbons also estimated that total operating costs for a typical U.K. malthouse comprise 25–30% fuel, 15–20% electricity, 15–25% wages, 10–20% repairs and maintenance, and 15–25% miscellaneous.

Fig. 76 (pg 166, above): Two-hundred kilogram pilot malthouse at Canadian Malt Barley Technical Centre Winnipeg, Canada, manufactured by MacDonald Steel Ltd. Cambridge, Ontario. Fig. 77 (pg 166, below): Giracleur steep-leveling and unloading machine.

STEEP-OUT, OR CASTING The point at which steeping is complete and the grain is ready to be transferred to another vessel for the germination cycle is called steep-out or casting. Target steep-out moisture is around 42–48% with a uniform **chitting** percentage of >96%.

In floor maltings, steep-out may occur by shoveling out of the cistern into the couch or pile on the flat germination floor.

Wet steep-out, or wet casting, by hydraulic transfer, means the steeped grains and final steep soak are dropped together in a slurry either by gravity or by pumping into the germination chamber. Due to increased hydraulic pressure, wet steep-out has been shown to produce a respiration lag in barley of as much as one half to a full day, as measured by a respirometer device. Certain enzyme activities and extract fermentability levels have also been reported to be reduced by wet steep-out methods (Yoshida, 1979).

Dry steep-out involves the use of mechanical arms or rakes, like the Giracleur machine developed in France, or conveyors to remove steeped barley from the tank. The Giracleur is a set of arms suspended from a central shaft in large steep tanks used to level the grain during steep-filling and push steeped grains out through steep-emptying valves located at the periphery of steep tank walls, in a "dry steep-out" process. Care must be taken to adjust these mechanical devices properly, otherwise physical mistreatment has been known to cut off emerged rootlets or damage protective layers of husk, leading to uneven germination and higher malt losses.

Grain beds must be leveled soon after steep-out for uniform germination and modification. Malthouse with unattended or wet steep-outs generally produce several piles of grain in the germination compartment. These piles should not take more than two trips of the turning machine or 12 hours to level completely.

To avoid any temperature shock and respiration lag at steep-out, it is sensible to match final steep grain and transfer water (if steep-out is wet) temperature with initial germination compartment temperature (60–66°F, 16–19°C).

Fig. 78 (above): Steep-out at Michigan Malting Co., Shepherd, MI.

Fig. 79 (pg 168 below): Blacklands Malt (Austin, TX) cylindroconical steep tank, Germinate-Kiln vessel (GKV) and air heating and spray chambers.

STEWING A process used by maltsters to create specialty caramel or crystal malts. Stewing involves holding **green malt** while still wet at high temperatures, causing the endosperm to liquefy and crystalize when dried, producing sweet, caramel or toffee-like flavors.

STORAGE OF UNMALTED AND MALTED GRAINS During storage, grain moisture always moves toward equilibrium with the moisture content of the ambient air. Given enough time, ambient air relative humidity of 50% will cause grain moisture to equalize at 12% moisture, whereas extended periods of rainy, wet ambient air at 90% relative humidity will equalize at 21% grain moisture. Drawing cooler night-time air, which is lower in relative humidity, through the grain, is a good practice to slow the transition of water from air to grain in flat or bin storage.

STRIPE RUST (Puccinia striiformis) A plant fungus that can produce orange pustule stripes along the leaf, impeding photosynthesis and crop yield. Stripe rust resistance is a heritable trait, varying among varieties.

STUCK MASH When mash run-off (transfer from mash or lauter tun to brewkettle) stops or is severely delayed in the brewhouse, sometimes taking 10–20 hours, rather than the 1–3 hour normal runoff cycle. There are many equipment, malt, and brewing process-related causes of stuck mashes. The malt-related causes can be: a malt grind that is too fine, producing small husk particles that interferes with good mash bed development and extract flow; under-modified malt or malt with a high proportion (>2%) of dead grains; insufficient malt enzyme activity to total (malt + adjuncts) starch substrate ratio; or malt that has gone slack due to excessively long or humid storage conditions.

SULFUR DIOXIDE Burning elemental sulfur in direct-fired kilns is still used in some malthouses to control nitrosamine levels in finished malts. High sulfur dioxide levels in malt can reduce wort pH and enzyme activity rates necessitating longer fermentation times.

Sulfur dioxide can also lighten the color and reduce microflora on the surface of malt. Oxides of sulfur may be created in direct kilning which produce black spots on the surface of malt, called the "magpie" effect.

SUSTAINABILITY OF CEREAL GRAIN SUPPLY TO MALTING Sustainability of supply of high-quality grains for malting depends on many agronomic and business factors, including: competition for acreage from feed barley; production of biofuels, or cereals for starch-based packaging; and increasing grain yield per acre, resulting in fewer total acres planted.

An estimate of the U.K. malting industry by the Maltsters Association of Great Britain (MAGB) showed that some 60% of malt's total carbon footprint occurs in crop growing. Nitrogen fertilizers account for 48%, emission of nitrogen oxides (NOx) from soil (35%), and fuels (17%), of malt's total carbon footprint. Heating gas adds another 25% to the total carbon footprint, electrical power 10%, and 5% each from barley and malt storage, transportation, and water/wastewater supply and treatment.

Nigel Davies of Muntons plc (Suffolk, England) estimated that up to 75% of the carbon footprint of malt can be reduced. Some ideas put forward in this discussion include "greener" and more efficient application of fertilizers, lowering kilning temperatures to yield base malts at 6% rather than 4% final moisture, introduction of mandatory NOx scrubbing, and use of food wastes and sludge from wastewater digesters as crop fertilizers (Davies, 2010).

Later, Davies (2013) outlined the potential for worldwide malthouses to reduce average water usage by nearly half, from 5.5 m3/ton (1,452 gal/ton) to <2.5 m3/ton (660 gal/ton) and energy usage from 850 kWh/ton to <500 kWh/ton.

T

TERROIR The word "*terroir*" is French for "local" and loosely translates as "a sense of place." Terroir is used as marketing leverage in the wine, coffee, chocolate, tea, and even cannabis industries, to characterize the impact of geography, geology, latitude, longitude, soil type, and rainfall on product quality, brand identity and mystique. In extreme cases, terroir has legal footing in Protected Appellations of Origin, or Appellation d'origine contrôlee (AOC) system for things like wine and cheese

As Professor Charles Bamforth points out in his book *Grape vs. Grain* (Cambridge Press, 2008), "Beers might fairly be marketed on the basis of hop varietal just as much as is a wine on a grape varietal. Invariably, they aren't." Perhaps things are starting to change with new or heritage grain varieties, individual farms, as well as malthouse size and location being added to the list of product information desired by today's (and tomorrow's) increasingly savvy consumers.

THERMOMETERS (FROM BRANDON ADE, BLACKLANDS MALT LLC, AUSTIN, TX) "No two digital or analog meters will give you the same reading. So which do you trust and what do you do? Answer: You need to get a calibrated and certified reference standard by which all your other devices are calibrated from.

"In my experience digital meters suck. They break all the time, really accurate and certified ones are expensive, and I just don't find myself trusting their readings or reliability. So, find yourself a good liquid-filled glass thermometer. That's right, the constant thermodynamic properties of liquids beat the tar out of the digital silicon age in reliability and affordability.

"Mercury filled thermometers are nearly impossible to find these days as the government issued an order a few years ago to phase them out due to dangers of poisoning. But there are a variety of organic liquid (mineral spirits, kerosene, toluene) filled ones on the market. You can easily recognize them by dyed colors of green, blue, red, black. Mercury thermometers are of course filled with silver.

"They can be expensive, $200–500. But what you are buying is assurance and peace of mind that the temperature you are reading is both precise and accurate. You can then calibrate your other equipment off of your reference standard thermometer. It's worth it to know your germination and kiln temperatures are accurate!

"Here are some general tips when searching for a good thermometer:
- Total immersion is more accurate than partial immersion.
- Teflon encapsulated thermometers are safer if the glass breaks, but are less precise over time.
- Yellow glass back with black/dark liquid are easiest to read.
- Look for "NIST traceable factory certified," "Spirit filled," "total immersion."
- Some brands are H-B and Thermco.
- Make sure the thermometer comes with a calibration report that indicates how to correct your readings based on temperature range.
- Make sure the calibration report meets the current test methods. As of this writing:
- ASTM E-77, ASTM E1, ANSI/NCSL Z540.3-2006, or NIST SP 1088
- Don't trust anything else that is not certified and calibrated!"

THOUSAND KERNEL WEIGHT (TKW) The weight of 1,000 kernels of any grain on a dry basis, after screening, cleaning, and removal of foreign grains. TKW gives an indication of grain plumpness and sizing, useful when comparing crop years or varieties.

TILDEN DRUM MALTING SYSTEM An automated drum germination and kilning system developed in 1910 made of two concentrically mounted perforated metal cylinders between which steeped grains were loaded and allowed to germinate while rotating. Humidified and cooled air was passed through the cylinders and grain during germination and replaced with dry, heated air during kilning.

TOWER MALTING   A vertically arranged malthouse that uses gravity, as much as feasible, to convey grains downward from one malting step to the other and minimize foundation and roofing construction costs. Steep tanks are located on the top floor (sometimes hundreds of feet high), and are steeped-in by bucket elevators from rail or truck unloading stations at ground level through an adjoining utility tower.

Steep-outs may be wet (using water and gravity to convey the chitted malt down to germination), but risk of hydraulic shock and respiration lag increases with the height of vertical drop, therefore, tower malthouse steep-outs are usually dry using leveling and emptying arms such as the **Giracleur steeping technology**.

Germination vessels, usually circular like the tower, are replicated on several floors below the steep level. For two identical steep tanks, a total of six or seven germination compartments can be built in the tower. Ducts for air conditioning, spray chambers, fresh and recirculated air valves are usually located in the adjacent utility tower.

Circular kilns may be located in the same tower or in an adjoining building along with malt cleaning equipment.

Circular germination compartments and kilns have been built with rotating turning machines and fixed floors or, conversely, rotating floors and fixed turning machines.

TRITICALE   *(triticosecale)*   Triticale is a hybrid between wheat (*Triticum*) and rye (*Secale*). Triticale may be malted, producing high diastatic power (DP) malts, or can be used as a non-malted adjunct.

U

UNIVERSITY OF NOTTINGHAM CENTRE FOR BIOENERGY AND BREWING SCIENCE University in the midlands of the U.K. for teaching and research in brewing and bioenergy. (website www.nottingham.ac.uk/brewingscience).

V

VIENNA MALT Vienna malts are well-modified two-rowed barley malts produced in a conventional malt kiln. Continued germination and development of amino acids and sugars is promoted during the early stages of kilning by drying with cool air (100–120°F, 40–50°C) for several hours then slowly increasing temperature to 200–220°F (95–105°C) to keep malt color low.

Vienna malts have 4–5% moisture, and color of 2–5 SRM (4–10 EBC) and at least 80% extract, retaining enough enzyme activity by gentle kilning that they can be used as a base malt when called for.

VISCOAMYLOGRAPH A laboratory instrument used primarily in baking applications, but also used in malt analysis to show the reduction of viscosity produced by amylolytic enzymes of malt (Glennie Holmes, 1995). Viscoamylographs have been used to track the water migration and enzyme-susceptibility changes that occur during post-kiln aging of malt. A change in viscoamylogram characteristics has been observed around 21 days in pale lager malt aged in dry conditions at room temperature. Further work is needed using instruments like the viscomylograph to evaluate changes during malt aging.

VOLUME FORMULAS IMPORTANT TO MALTSTERS.

Rectangle or square	Volume = length × width × height
Cylinder	Volume = π × height × radius^2
Cone	Volume = (π × height × radius^2) /3
Pyramid	Volume = (length × width × height)/3

WATER Water used to initiate the malting process is added via immersion, or rarely, sprays, with grains to be malted. Steep water must be free of chlorine, microorganisms, herbicides, pesticides, and iron which can inhibit enzymes and taint malt flavors. Two, three or even four immersions with fresh, temperature-controlled steep water are used depending on water cost, availability and customer specifications. Immersion water temperature is usually coolest (50–60°F, 10–16°C) in the first immersion but, where possible, can be warmed to 60–66°F (16–19°C) in later soaks.

Adding slaked (hydrated) lime (calcium hydroxide, Ca(OH)2) to initial steep water is sometimes done to raise pH, which reduces microbial loading and leaches **polyphenols (tannins)** from the husk to improve colloidal (haze) stability of beer.

Reuse of steep water using vibrating screen filtration to remove particulates and biological oxygen demand (BOD) and reoxygenation is gaining increasing interest and affordability worldwide. However, one of the major hurdles is the fact that recycled steep water can be made potable, but there remain parts-per-million quantities of compounds that inhibit grain germination and slow malt modification.

In 2005, a consortium of French and U.K. maltsters obtained grants from their respective governments for a joint malthouse water-saving project called SWAN (Save Water Attend Nature). Their goal was to reuse up to 85% recycled water in each steep. These researchers were able to identify the germination inhibition compounds as quinone oxidation products and developed and effective steep water treatment system using membrane bioreactors coupled to reverse osmosis. The system as designed (Davies, 2009) was able to recycle up to 70% of steep water and meet quality goals, showing ROI (return-on-investment) of as little as 1.3 years for malthouses that purchase water from local authorities and discharge (and pay for) wastewater to local sewers.

Water usage in malthouses is about 700–1,000 gallons per ton of malted barley, which can then make about 35 barrels of all-malt "craft" or 70 barrels of U.S. premium or light lagers brewed with added non-malt adjuncts and generally packaged at lower alcohol levels (not including water added in brewing), or 150 gallons of 80 proof (40% abv) double-distilled whisky (Owens, 2008). Efforts are underway by breeders in Australia, suffering multi-year droughts, to breed varieties of barley that are lower in, or have negligible amounts of ß-glucan, thereby requiring single water steeps, rather than three or four commonly used (Stewart, 2009).

WHEAT *(Triticum aestivum)* Wheat is the most widely grown cereal crop worldwide. Because harvested wheat has no husk adhering to it, water uptake in steeping is more rapid for wheat than barley of the same size. Rootlet and acrospire growth is also more vigorous than barley, causing more clumping in germination and is much more susceptible to damage in turning. Lower steep-out moistures for wheat (35–40%) in 24 hour, two-immersion steep cycles are advisable with spray additions in germination to bring moisture to 40–42%.

Wheat can be germinated and kilned at lower temperatures than barley. Germination temperatures of 50–60°F (10–15°C) for three or four days and final kiln curing temperatures 165–180°F (74–81°C) may be sufficient for pale wheat malts with a color of 1–2 SRM (2–4 EBC). Dark wheat malt is cured at 212–230°F (100–110°C) to produce colors of 7–10 SRM (14–20 EBC).

The lack of husk also makes it easier to damage by germination turning machines, grain elevators, and grain plows in the malthouse.

Malted wheat is used for unique non-barley flavors and

improved head retention in beer and can replace malted barley up to about 70% in brewhouses and distilleries with traditional lauter tuns. High levels of pentosan in wheat can cause beer haze. Wheat starch gelatinizes at lower temperatures, 125–147°F (52–64°C), than barley starches (145–154°F, 63–68°C).

Additional mash filtration aids like rice hulls are sometimes added to mashes with a high proportion of malted wheat. Breweries with mash filters, rather than lauter tuns, have been able to brew with 100% malted wheat in products, sometimes combined with wine grapes, called "wheat wines."

Malted wheat is used in making Munich Weissbier, and the fermented foods *bouza, talla, and kishk* in Africa, and *chan* and *takju* in Asia.

Wheat flour is used up to 10% of the malt grist in beer to reduce cost, lower protein levels, and improve beer head retention.

WINTER BARLEY A winter barley is planted in late fall and is harvested 9 to 10 months later in the following summer. If a winter barley is planted late, say the same time as spring barley, it will not flower or will flower late, drastically reducing yield. There are many more spring barleys than winter barleys with malting quality in the U.S., although there is renewed interest in increasing winter barley production.

The American Malting Barley Association (AMBA) issued this press release on February 19, 2014 regarding the increased interest in winter barley production in the U.S.:

> "Barley acreage in the U.S. has declined to levels not seen in over 100 years, but the interest in growing malting barley has spread to regions in the U.S. that have not grown the crop for many decades. The driving force behind this interest is the movement to source food locally. In some cases, the movement has been aided by legislation. In New York for example, a special Farm Brewery License can be obtained for beers that are made from primarily locally sourced ingredients. The desire to source

and process locally produced malting barley brings many opportunities as well as some challenges. Barley is an excellent small grain in rotation with any number of crops. In addition to breaking disease cycles, barley's fine root structure helps condition soils and its early maturity allows for establishment of fall seeded [winter] crops or even double cropping in some regions. With relatively low inputs and added premiums, malting barley is a very attractive crop option."

According to Mike Davis of AMBA,
"Winter varieties, which are planted in the fall, can yield up to 20% more than spring varieties and require one less watering when grown under irrigation, an important resource and cost consideration for growers. Winter hardiness is the primary restraint that needs to be addressed to make winter barley competitive with winter wheat."

Even American craft brewers are getting involved in the promotion of winter barleys. John Bryce, then of Blacksburg Brewery in Virginia, read a paper entitled "Should Winter Barley Breeding be Intensified?" at the 2009 annual meeting of the MBAA at La Quinta, CA, presenting data he and his colleague Felipe Alvisi Galastro Perez obtained from micro-malting and micro-brewing trials of a new experimental winter barley at the Versuchs-und Lehranstalt für Brauerei (VLB) in Berlin.

In 2006, several factors (reduced acreage, bad weather, and poor crop quality) led to a one million tonne shortfall in high-quality spring two-rowed malting barley harvested in Germany, forcing a significant increase in malting and brewing with two- and six-rowed winter barleys. Niemsch (2007) reported on large-scale brewing trials of lower quality winter/spring barley malt blends (50% winter/50% spring) compared to the control 100% spring barley malt brews. The 50% two-rowed winter barley malt blend test brews showed 2.9% lower wort fermentation attenuation, 7 ppm lower FAN, 75 ppm higher wort ß-glucan, 0.08 mPas higher wort viscosity, 0.10 higher beer pH, and 17 ppm higher beer ß-glucan than the 100% spring barley malt control. The other 50% winter barley replacement brews (30% six-rowed winter and 20% two-rowed winter barleys) produced even lower quality wort and beer, with 5.8% lower wort fermentation attenuation, 32 ppm lower wort FAN, 268 ppm higher wort ß-glucan, 0.39 mPas higher wort viscosity, 0.25 higher beer pH, and 159 ppm higher beer ß-glucan. The author further reported on trials using silica gel for haze-protein removal and polyvinylpoly-pyrrolidone (PVPP) for tannin adsorption to improve final beer colloidal (haze) stability in the winter barley malt blend brews.

WITHERING Withering is an extra step between germination and kilning which is largely archaic, but can be performed in today's pneumatic by turning down airflow at the end of germination, prior to transferring to the kiln. Sir Robert Moray (1678) said:

"As soon as the malt is sufficiently come, turn it over, and spread it to a depth not exceeding five or six inches, and by the time it is all spread out, turn it over and over three or four times, after this turn it once in four or five hours, making the heap thicker by degrees, and continuing to do so constantly for the space of 48 hours at least; this frequent turning it, cools, dries, and deadens the grain, whereby it becomes mellow, melts easily in brewing, and then separates entirely from the husk; after which throw up the malt into a heap, as high as you can; where let it lie till it grow as hot as your hand can bear it, which usually happens in about 30 hours; and this compleats the sweetness and mellowness of the malt; after the malt is sufficiently heated, spread it to cool, and turn it over again in six or eight hours after; then dry it upon a kiln."

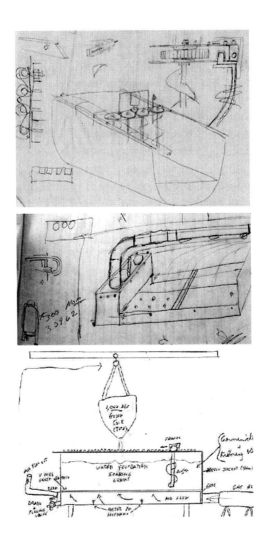

Figs. 80–82: The beginnings of dreams coming true?

181

REFERENCES

American Society of Brewing Chemists Methods of Analysis (14th edition), Scientific Societies, St. Paul, MN.

Armstrong, H. E. (1934) *Journal of the Institute of Brewing.* **40**, 223–225.

Bamforth, C. W. (2002) Nutritional aspects of beer—A review. *Nutrition Research*, **22**, 227–237.

Bamforth, C. W. and Barclay, A. H. P (1993) Steeping—The crucial factor in determining malt quality. *Brewers Digest*, **August**, 20–23.

Barnard, A. (1889) *The Notable Breweries of Great Britain and Ireland.* Causton & Sons, London.

Bathgate, G.N. (1973) The biochemistry of malt kilning. *Brewer's Digest*, **48**, 60–65.

Beattie, S., Schwarz, P., Horsley, R., Barr, J. and Casper, H. (1998) The effect of grain storage conditions on the viability of *Fusarium* and Deoxynivalenol production in infested malting barley. *Journal of Food Protection*, **62**, 103–106.

Bell, D. (2011) *Alt Whiskeys: Alternative Whiskeys and Techniques for the Adventurous Distiller.* (A. Bell, ed.), pp. 144–165, American Distilling Institute, Hayward, CA.

Boulton, C. (2013) *Encyclopaedia of brewing.* Wiley-Blackwell, West Sussex, U.K.

Brewers Association (2014) Malting Barley Characteristics for Craft Brewers white paper. Available online at www.brewersassociation.org/attachments/0001/4752/Malting_Barley_Characteristics_For_Craft_Brewers.pdf.

Brigg, D. E. (1998) *Malts and Malting*, pp. 339, 466, Blackie, London.

Brookes, P. A., Lovett, D. A., and MacWilliam, I. C. (1976) The steeping of barley, A review of the metabolic consequences of water uptake, and their practical implications. *Journal of the Institute of Brewing*, **82**, 14–26.

Brown, J. (1983) *Steeped in Tradition: The malting industry in England since the railway age.* Institute of Agricultural History, University of Reading, U.K.

Brown, H. T and Morris, G. H. (1890) *Journal Chemical Society,* **57**, 458–528.

Brown, A., McWilliam, B. and Kinsman, M. (2011) *Brewer & Distiller International.* **7**(9), 55–58.

Chapon, L., Erber, H.-L., Kretschmer, K.-F., and Kretschmer, H. (1977) *Monatsschrift für Brauerei,* **32**, 160.

Clark, C. (1998) *The British Malting Industry since 1830.* Hambledon Press, London.

Colgrave, M. L., Goswami, H., Howitt, C. A., and Tanner, G. J. (2013) Proteomics as a tool to understand the complexity of beer. *Food Research International,* **54**, 1001–1012.

Darby, W. J., Ghalioungi, P., and Grivetti, L. (1977) *Food: The gift of Osiris.* Academic Press, London.

Davies, N. (2009) Sustainable malting, recycling process water. *Brewer & Distiller International,* **5(4)**, 34–37.

Davies, N. (2010), Greening the malt supply chain. *Brewer & Distiller International,* **6**, 50–53.

Davies, N. (2013) Low carbon malt—The journey continues. *Brewer & Distiller International,* **9**, 12-15.

Dolan, T. C. S. (2000) Scotch malt whisky distillers malted barley specifications: Twenty years of fermentable extract and predicted spirit yield. *Journal of the Institute of Brewing,* **106**, 245–249.

Dolan, T. C. S. (2003) Malt whiskies: Raw materials and processing, Chapter 2. *Handbook of Alcoholic Beverages Series, Whisky: Technology, Production and Marketing* (I. Russell, ed.), pp. 27–74, Elsevier Ltd., NY.

Evans, D. E. (2012) The Impact of Malt Blending on Lautering Efficiency, Extract Yield, and Wort Fermentability. *Journal of the American Society of Brewing Chemists,* **70**, 50–54.

Giarratano, C. and Thomas, D. (1986) Rapid Malt Modification

Analyses in a Production Malthouse: Friabilimeter and Calcofluor Methodologies. *Journal of the American Society of Brewing Chemists*, **44**, 95–97.

Gibson, G. (1989) in *Cereal Science and Technology* (G. H. Palmer, ed.), p. 279, Aberdeen University Press, U.K.

Glennie Holmes, M. (1995) Studies on Barley and Malt with the Rapid Viscoanalyzer. *Journal of the Institute of Brewing*, **101**, 19–28.

Guerdrum, L. J. and Bamforth C. W. (2012) Prolamin Levels Through Brewing and the Impact of Prolyl Endoproteinase. *Journal of the American Society of Brewing Chemists*, **70**, 35–38.

Hartong, B. D. and Kretschmer, K. F. (1958) *Monatsschrift für Brauerei*, **11**, 238–242.

Hartong, B. D. and Kretschmer, K. F. (1969) *Brauwelt*, **109**, 1785–1788.

Hertrich, J. (2013) *Master Brewers Association of the Americas Technical Quarterly*, **50**, 131–141.

Hopkins, R. H. and Krause, C. B. (1937) *Biochemistry Applied to Malting and Brewing*, pp. 130–136, Allen and Unwin, London.

Hornsey, I. (2003) *A History of Beer and Brewing*, The Royal Society of Chemistry, Cambridge, U.K.

Hough, J. S., Briggs, D., and Stevens, R. (1971) *Malting and Brewing Science*, pp. 38–41, Chapman and Hall, London.

Hubner, F. and Arendt, E. (2010) Studies on the Influence of Germination Conditions on Protein Breakdown in Buckwheat and Oats. *Journal of the Institute of Brewing & Distilling*, **116**, 3–13.

Ishida, H. (2002) *Master Brewers Association of the Americas Technical Quarterly*, **39**, 81–88.

Kinsman, M. (2011) Scotland's newest maltings. *Brewer & Distiller International*, 5, 38–39.

Kirsop, B. H. and Pollock, J. R. A. (1958) *Journal of the Institute of*

Brewing, **64**, 227–233.

Kunze, W. (1999) *Technology Brewing and Malting*, 2nd edition, VLB Berlin.

Lekkas, C., Hill, A., and Stewart, G. G. (2014) Extraction of FAN from Malting Barley During Malting and Mashing. *Journal of the American Society of Brewing Chemists*, **72**, 6–11.

Li, Y. and M.-J. S. Maurice (2013) Development of a Fast and Reliable Microwave-based Assay for Measurement of Malt Color. *Journal of the American Society of Brewing Chemists*, **71**, 144–148.

MacLeod, A. M. (1969) *Science Progress*, **57**, 99–112, Oxford, U.K.

MacLeod, A. M. and Napier, J. P. (1959), *Journal of the Institute of Brewing*, **65**, 188–196.

Mauch, A., Wunderlich, S., Zarkow, M., Becker, T., Jacob, F., and Arendt, E. (2011) The Use of Malt Produced with 70% Less Malting Loss for Beer Production: Part I & II. *Journal of the American Society of Brewing Chemists*, **69**, 227–254.

Mayer, H., Marconi, O., Perreti, G., Sensidoni, M., and Fantozzi, P. (2011) Investigation of the Suitability of Hulled Wheats for Malting and Brewing. *Journal of the American Society of Brewing Chemists*, **69**, 116–120.

Moll, M. (1979) in *Brewing Science* (J. R. A. Pollock, ed.), pp. 539–575, Academic Press, London.

Moray, Sir Robert (1678) The making of Malt. *Philosophical Transactions*, 142, 1069–1072.

Niemsch, K. (2007) Brewing with chicken feed, a view from continental Europe on the 2006 barley crop. *Brewer & Distiller International*, **3**, 37–39.

Owens, B. (2008) *Craft Whiskey Distilling*, American Distilling Institute, Hayward, CA.

Palmer, G. H. (1967) Ph.D. Thesis, University of Edinburgh, U.K.

Palmer, G. H. (1973) *Journal of the Institute of Brewing*, **79**, 513–518.

Palmer, G. H. (1989) Cereals in Malting and Brewing, in *Cereal*

Science and Technology (G. Palmer, ed.), pp. 61–242.

Palmer, G. H. (1999) Achieving Homogeneity in Malting, in *Proceedings of the European Brewery Convention Congress*, pp. 232–263, Cannes, France.

Pauls Malt (1995–1997) Engineering Information, *Brewing Room Book*, pp. 271–272, Suffolk, U.K.

Pauls Malt (1998–2000) *Brewing Room Book*, p. 220 Suffolk, U.K.

Ponsonby, R. (2007).Appel's malt trials. *Brewer & Distiller International*, 3, 24–26.

Pyler, D. and Thomas, D. (2000) Malted Cereals: Their Production and Use. *Handbook of Cereal Science and Technology*, 2nd edition (Kulp and Ponte, eds.), pp. 685–696, Marcel Dekker, NY.

Ringrose, D. (1979) Energy conservation—The European maltsters efforts. *Master Brewers Association of the Americas Technical Quarterly*, 16(2), 68–72.

Scamell, G. and Colyer, F. (1880) *Breweries and Maltings: Their Arrangement Construction, Machinery and Plant*, White Mule Press, London.

Schwarz, P. and Horsley, R. (2012) A Comparison of North American Two-row and Six-row Malting Barley, MoreBeer, url http://morebeer.com/brewingtechniques/bmg/schwarzsb.html

Shelton, D. and Lee, W. J. (2000) Cereal Carbohydrates. *Handbook of Cereal Science and Technology*, 2nd edition (Kulp and Ponte, eds.), pp. 385–415. Marcel Dekker, NY.

Stevens, P. (1993) *Newark the Magic of Malt*, Nottinghamshire, U.K.

Stewart, D. (2009) Beating beta-glucan, the key to maltings water savings. *Brewer & Distiller International*, 5(7), 39.

Stopes, H. (1885) *Malt and Malting*, Lyon, London.

Tanner, G., Colgrave, M., Howitt, C. (2014) *Journal of the American Society of Brewing Chemists*, 72(1), 36–50.

Tawabe, J.-C. B, Bera, F., and Thonart, P. (2012) Optimizing red sorghum malt quality when Bacillus subtilis is used during steeping to control mould growth. *Journal of the Institute of Brewing*, **118**, 295–304.

Thomas, D. (1983) *Studies on the Enzymic Hydrolysis of Barley Endosperm during Malting*, M.Sc. Thesis, Heriot-Watt University, Edinburgh.

Thomas, D. (1986) A novel result of malt friabilimeter analysis: Case-hardened malt. *Journal of the Institute of Brewing*, **92**, 65–68.

Thomas, D., Bright, D., and Irving, J. (1990) Aspects of germination in some new barley varieties. *Proceedings of the Third Aviemore Conference on Malting, Brewing & Distilling* (I. Campbell, ed.), pp. 371–374, Institute of Brewing, London.

Thomas, D., and Palmer, G. (1994) Malt. *The New Brewer*, Mar–Apr.

Tryon, T. (1691) in *The New Art of Brewing Beer, Ale and Other Sorts of Liquors*, pp. 45–57, Salisbury, London.

Wahl, R. and Henius, M. (1908) in *American Handy Book of the Brewing, Malting and Auxiliary Trades*, 3rd edition, Wahl-Henius Institute, Chicago.

Walker, E. W. (1990) The application of peat smoke to distilling malt. In *Proceedings of the Third Aviemore Conference on Malting, Brewing and Distilling* (I. Campbell, ed.) pp. 348–355. Institute of Brewing, London.

Wellington, P. A. (1965) in *The Growth of Cereals and Grasses* (F. L. Milthorpe and J. D. Ivins, eds.), pp. 3–19, Butterworths, London.

Wentz, M., Horsley, R., and Schwarz, P. (2004) Relationships among Common Malt Quality and Modification Parameters, *Journal of the American Society of Brewing Chemists*, 62, 103–107.

White, E. S. (1860) *The Maltster's Guide*, Loftus, London.

Yoshida, T., Yamada, K., Fujino, S. and Koumegawa, J. (1979) *Journal of the American Society of Brewing Chemists*, **37**, 77.

APPENDIX A

THE MALTING PROCESS FLOW OF HEAT, AIR, GRAIN AND WATER (FROM PYLER AND THOMAS, 2000)

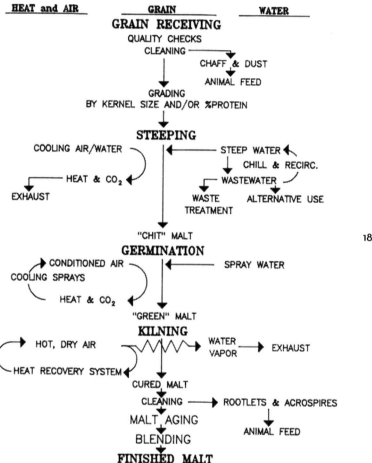

APPENDIX B
RECOMMENDED MALT CHARACTERISTICS
AMERICAN MALTING BARLEY ASSOCIATION

American Malting Barley Association, Inc.
MALTING BARLEY BREEDING GUIDELINES
IDEAL COMMERCIAL MALT CRITERIA

	Six-Row	Adjunct Two-Row	All Malt Two-Row
Barley Factors			
Plump Kernels (on 6/64)	> 80%	> 90%	> 90%
Thin Kernels (thru 5/64)	< 3%	< 3%	< 3%
Germination (4ml 72 hr. GE)	> 98%	> 98%	> 98%
Protein	≤ 13.0%	≤ 13.0%	≤ 12.0%
Skinned & Broken Kernels	< 5%	< 5%	< 5%
Malt Factors			
Total Protein	≤ 12.8%	≤ 12.8%	≤ 11.8%
on 7/64 screen	> 60%	> 70%	> 75%
Measures of Malt Modification			
Beta-Glucan (ppm)	< 120	< 100	< 100
F/C Difference	< 1.2	< 1.2	< 1.2
Soluble/Total Protein*	42-47%	40-47%	38-45%
Turbidity (NTU)	< 10	< 10	< 10
Viscosity (absolute cp)	< 1.50	< 1.50	< 1.50
Congress Wort			
Soluble Protein*	5.2-5.7%	4.8-5.6%	< 5.3%
Extract (FG db)	> 79.0%	> 81.0%	> 81.0%
Color (°ASBC)	1.8-2.5	1.6-2.5	1.6-2.8
FAN	> 210	> 210	140-190
Malt Enzymes			
Diastatic Power (°ASBC)*	> 150	> 120	110-150
Alpha Amylase (DU)*	> 50	> 50	40-70

General Comments

Barley should mature rapidly, break dormancy quickly without pregermination and germinate uniformly.

The hull should be thin, bright and adhere tightly during harvesting, cleaning and malting.

Malted barley should exhibit a well-balanced, modification in a conventional malting schedule with four day germination.

Malted barley must provide desired beer flavor.

April, 2014 DRAFT

Reprinted with permission of the American Malting Barley Association, Milwaukee, WI.

APPENDIX C

NORTH AMERICAN CRAFT MALTING COMPANIES (AS OF APRIL, 2014)

NORTH AMERICAN CRAFT MALTERS	SALES (S) BREWERY (B) OR DISTILLERY (D) MALTHOUSE	MALTSTER	PHONE	LOCATION	STARTED MALTING	BATCH SIZE (TONS)	ANNUAL CAPACITY (TONS)	CONTACT
11 Wells Spirits	D	Bob McManus	651-300-9328	St. Paul, MN	construction	1	40	11wells.com
Academy Malt Co.	S	Jeremy Weaver	317-604-0771	Indianapolis, IN	Jan. 2013	1	80	
Blacklands Malt	S	Brandon Ade	530-289-6258	Austin, TX	2013	2	107	blacklandsmalt.com
Blue Ox Malthouse	S	Joel Alex	207-649-0018	Belfast, ME	2013			blueoxmalthouse.com
California Craft Malt Co.	S	Kevin Christensen & Ron Silberstein	415-974-0905	Oakland, CA	planning			
California Malting Co.	S	Curtis Davenport	805-325-9020	Santa Barbara, CA	construction	1	100	californiamaltingco.com
Christensen Farms Malting Co.	S	Zach Christensen	503-550-3576	McMinnville, OR	2009	0.5	68	heritagemalt.com
Colorado Malting Co.	S	Wayne, Josh and Jason Cody	719-580-5084	Alamosa, CO	2008	2.5	600	coloradomaltingcompany.com
Copper Fox Distillery	D	Rick Wasmund	540-987-8554	Sperryville, VA	construction			copperfox.biz/
Corsair Artisan Distillery	D	Darek Bell		Nashville, TN	2010	2	100	corsairartisan.com

NORTH AMERICAN CRAFT MALTERS	SALES (S) BREWERY (B) OR DISTILLERY (D) MALTHOUSE	MALTSTER	PHONE	LOCATION	STARTED MALTING	BATCH SIZE (TONS)	ANNUAL CAPACITY (TONS)	CONTACT
Deer Creek Malthouse	S	Mark Brault	717-746-6258	Glenn Mills, PA	construction	2	120	deercreekmalt.com
Doehnel Floor Malting	S	Mike Doehnel	360-982-1395	Victoria, BC, Canada	1998	1	11	
Eckert Malting & Brewing	S/B	Jim Eckert	530-230-7028	Chico, CA	2012	0.8	20	eckertmalting@gmail.com
Empire Malt Works	S	David VanHoute	347-628-2294	Albany, NY	construction	1	50	dvanhoute@empiremaltworks.com
Epiphany Craft Malt	S	Sebastian Wolfrum	919-699-6733	Durham, NC	construction	18	600	epiphanycraftmalt@gmail.com
Farm Boy Farms	S	Daniel Gridley	919-696-7850	Pittsboro, NC	2011	0.5	125	farmboybrewery.com
Farmhouse Malt NYC	S	Marty & Natlie Mattrazzo	607-227-0638	Newark Valley, NY	2013	1	50	farmhousemalt.com
Gold Rush Malt LLC	S	Tom Hutchison	541-240-1280	Baker City, OR	construction	2.5	125	goldrushmalt.com
Grouse Malting & Roasting Co.	S	Twila Henley	419-606-1852	Wellington, CO	2013	1	100	grouseco.com
High Desert Malt & Grain	S	Ed Galazzo	541-389-0486	Bend, OR	construction			
Hillrock Estate Distillery	D	Dave Pickerell	518-329-1023	Ancram, NY	2008	1	100	hillrockdistillery.com
Keystone Malt LLC	S	Alan Gladish, Gary Snyder	215-806-1619	Philadelphia, PA	planning	1.5	75	allan@keystonemalt.com
Leopold Bros. Distilling	D	Todd Leopold	720-938-7430	Denver, CO	construction	7	250	leopoldbros.com

NORTH AMERICAN CRAFT MALTERS	SALES (S) BREWERY (B) OR DISTILLERY (D) MALTHOUSE	MALTSTER	PHONE	LOCATION	STARTED MALTING	BATCH SIZE (TONS)	ANNUAL CAPACITY (TONS)	CONTACT
Maine Craft Distilling	D	Luk Davidson			construction			mainecraftdistilling.com
Maltarie Frontenac Inc.	S	Bruno Vachon	418-338-9563	Quebec, Canada	2006	5.5	825	malteriefrontenac.com
Mammouth Malt	S	Eric Steigman		Thawville, IL	2003	0.8	30	
Mecca Grade Estate Malt	S	Seth Klann	541-231-2801	Madras, OR	construction	10	750	meccagrade.com
Michigan Malt	S	Wendell Banks	989-954-5962	Shepherd, MI	2000	1	50	michiganmalt.com
Mountain Malting	S	Mark E. Gonsky		Bend, OR	construction	4	380	markgonsky@gmail.com mountainmalting.com
New York Craft Malt	S	Ted Hawley	585-813-5389	Batavia, NY	2013	1	156	newyorkcraftmalt.com
Niagara Malt	S	Bob Johnson	716-961-9887	Cambria, NY	2014	1	50	niagaramalt.com
Ohio Malting Co.	S	Bob Matus	440-774-9463	Wakeman, OH	construction			ohiomaltingcompany@gmail.com
Our Mutual Friend Malt & Brew	B	Bryan Leavelle	951-265-4192	Denver, CO	2012	0.1	1	omfmb.com
Owens & Sons Malt Co.	S	Erik Owens		Hayward, CA	planning			
Peterson Quality Malt	S	Andrew Peterson	802-989-0014	Monkton, VT	construction	2	200	qualitymalt@gmail.com
Pioneer Malting	S	Adam Filippetti			planning			
Pilot Malt House	S	Eric May	616-209-8388	Jenison, MI	2012	1	30	pilotmalthouse.com

NORTH AMERICAN CRAFT MALTERS	SALES (S) BREWERY (B) OR DISTILLERY (D) MALTHOUSE	MALTSTER	PHONE	LOCATION	STARTED MALTING	BATCH SIZE (TONS)	ANNUAL CAPACITY (TONS)	CONTACT
Rebel Malting Co.	S	Lance Jergenen	775-997-6411	Reno, NV	2004	1	40	rebelmalting.com
Riverbend Malt House	S	Brent Manning	828-450-1081	Asheville, NC	2012	4	200	riverbendmalt.com
Rocky Mountain Distilling	D	Brent Kallevig	970-306-7766		planning			
Rogue Ales Farmstead Malthouse	B	Eric Hyatt		Newport, OR	2013	0.4	15	rogue.com
Skagit Valley Malting	S	Mike Doehnel & Wayne Carpenter	360-391-5492	Mt. Vernon, WA	construction	2	500	wcarpenter@skagitvalleymalting.com
Slow Hand Malting	S	David Morgan	802-846-7681	Hinesburg, VT	construction			slowhandmalting.com
Sprague Farm & Brew Works	B	Brian Sprague	814-398-2885	Venango, PA	2011	0.1	10	kegofsprague@coaxpa.com
Troubador Maltings LLC	S	Chris Schooley, Steve Clark	773-704-4407 970-443-5958	Ft. Collins, CO	planning			troubadormaltings.com
Valley Malt	S	Andrea Stanley	413-349-9098	Hadley, MA	2010	1	300	valleymalt.com
Western Feedstock Technologies	S	Tom Blake	406-599-4889	Bozeman, MT	2014	0.1	20	blake@montana.edu

APPENDIX D

BARLEY TO BEVERAGE IN A BALLAD

"John Barleycorn" is slang for barley and any beverage made from malted barley (ale, lager, porter, stout, whiskey). The ballad was written by the Scottish poet, and erstwhile barley farmer Robert Burns in 1782, describing the sowing, harvesting, drying, malting, milling, and mashing of barley, ending with the enjoyment of the drink. Burns wrote many other memorable songs and poems, including the iconic New Year's Eve song "Auld Lang Syne."

"John Barleycorn," by Robert Burns

(Malting & brewing reference)

There was three kings into the east,   *(farmer,*
Three kings both great and high,   *maltster,*
And they hae sworn a solemn oath   *& brewer)*
John Barleycorn should die.

They took a plough and plough'd him down,
Put clods upon his head,
And they hae sworn a solemn oath   *(sowing barley*
John Barleycorn was dead.   *seed)*

But the cheerful Spring came kindly on,
And show'rs began to fall;
John Barleycorn got up again,
And sore surpris'd them all.   *(sprouting in field)*

The sultry suns of Summer came,   *(barley awns*
And he grew thick and strong;   *protect grain*
His head weel arm'd wi' pointed spears,   *from birds)*
That no one should him wrong.

The sober Autumn enter'd mild,
When he grew wan and pale;   *(grain ripening)*
His bending joints and drooping head
Show'd he began to fail.

His colour sicken'd more and more,   *(turning from*
He faded into age;   *green to brown)*
And then his enemies began
To show their deadly rage.

They've taen a weapon, long and sharp,   *(harvesting stalks*
And cut him by the knee;   *with sickle)*
Then tied him fast upon a cart,
Like a rogue for forgerie.   *(bundling)*

They laid him down upon his back,   *(threshing grain*
And cudgell'd him full sore;   *from stalks)*

They hung him up before the storm, *(drying in the sun)*
And turn'd him o'er and o'er.

They laid him out upon the floor, *(steeping and germination)*
To work him further woe;
And still, as signs of life appear'd, *(turning to prevent spoilage and mold)*
They toss'd him to and fro.

They wasted, o'er a scorching flame, *(kilning to dry and roast malt)*
The marrow of his bones;
But a miller us'd him worst of all, *(milling the dry malt for brewing)*
For he crush'd him between two stones.

John Barleycorn was a hero bold,
Of noble enterprise;
For if you do but taste his blood,
'Twill make your courage rise.

'Twill make a man forget his woe;
'Twill heighten all his joy;
'Twill make the widow's heart to sing,
Tho' the tear were in her eye.

Then let us toast John Barleycorn,
Each man a glass in hand;
And may his great posterity
Ne'er fail in old Scotland!

—Robert Burns, "Scotland's Favourite Son," (1759–1796)

CPSIA information can be obtained at www.ICGtesting.com
Printed in the USA
LVOW11*2229120315

430367LV00005B/9/P

9 780991 043620